Rare Animals and Plants of Haleakalā National Park

Arthur C. Medeiros
and
Lloyd L. Loope

Research Division,
Haleakalā
National Park

Illustrated by
Nanci Sidaras

CONTENTS

I. OVERVIEW

This booklet describes some of the rare native plants and animals found in Haleakalā National Park, the spectacular end products of millions of years of evolution in isolation. As one noted scientist observed, they are "marvels of creation." The plants and animals of the park are an important part of the natural heritage of Hawai'i and the United States. We have attempted to provide information in this booklet using prose understandable to the educated reader while reflecting the most accurate information available. A glossary of scientific terms is included to assist with terms that may not be entirely familiar to the non-biologist, and recommendations are given for further reading on the natural history of Hawai'i.

The Hawaiian archipelago comprises a southeast-trending chain of eight major and 124 smaller islands, reefs and shoals stretching 1600 miles across a highly isolated area of the northeast Pacific Ocean. The substantial age of the island chain (over 40 million years), its isolation (over 2000 miles from the nearest continent), and the variety of habitats created by Hawai'i's relatively high mountains has created a special biota of plants and animals, different from any other on earth. The island of Maui, roughly figure-eight shaped, was built by two principal volcanoes. The older West Maui volcano is heavily eroded and is about two million years old. Haleakalā, the East Maui volcano where the park is located, is considerably larger, less eroded, and about one million years old.

Haleakalā National Park encompasses 44 square miles from sea level to 10,023 feet in elevation for the perpetuation of native Hawaiian biota. The park, an internationally recognized Biosphere Reserve, includes the eroded summit crater of the dormant Haleakalā volcano, the high-elevation outer western slopes, and the deep and remote rain forests of Kīpahulu Valley down to the coast. For many rare Hawaiian plants and animals, some described in this booklet, Haleakalā National Park and adjacent natural areas provide a final chance for long-term survival.

Like many isolated islands, the Hawaiian archipelago is today plagued by foreign species of plants and animals introduced through the actions of humans, sometimes purposefully, other times accidentally. These so-called alien species have reduced populations of native species, especially endemic species, and in some cases threaten their immediate survival. The term "endemic" is used to describe the quality of being found only in a given locality or region, such as endemic to Maui or to the Hawaiian Islands. Hawai'i has one of the highest percentages of endemic species in the world. Although Haleakalā National Park is generally recognized as one of the most intact natural areas in the Hawaiian Islands, one-third of the park's plants and animals are alien species, some presenting serious problems for native species. Alien animals such as wild pigs, goats, mongooses, rats, Argentine ants, and western yellowjackets threaten the native biota. Alien plants such as strawberry guava, blackberry, kikuyu grass, and Hilo grass also present problems. The resources management staff of Haleakalā National Park and volunteers control some of the worst of the invaders. Nevertheless, the survival of many formerly abundant native species still remains in jeopardy.

II. MAMMALS

Vast expanses of Pacific Ocean prevented most mammals from establishing populations in Hawai'i before the arrival of humans. Only two native mammals are found in the State — the Hawaiian monk seal and the Hawaiian hoary bat.

Though the Hawaiian monk seal *(Monachus schauinslandi)* may have long ago inhabited the rocky coasts of Haleakalā, the species no longer regularly breeds on the shores of the main Hawaiian Islands. The Hawaiian seal is one of three monk fur seals in the world. The Caribbean species is now extinct. The Mediterranean species is endangered and very rare, like its Hawaiian relative. The name of this seal, *'īlio-holo-i-ka-uaua*, means "dog running in the roughness," and refers to the appearance of the seal in its often harsh ocean home. The seals are now largely restricted to the Northwestern Hawaiian Islands, a chain of small, primarily uninhabited, low islands and atolls stretching over 1200 miles northwestward from the coast of Kaua'i. Beginning in the early 1990s, for the first time in decades there have been increasing sightings of Hawaiian monk seals off Maui's coasts; perhaps more will occur if protection for this unique marine mammal continues. Never common in modern times, approximately 1200-1300 *'īlio-holo-i-ka-uaua* remain.

'ŌPE'APE'A

Hawaiian hoary bat
Lasiurus cinereus semotus
Common bat family (Vespertilionidae)

The other native Hawaiian mammal is the Hawaiian hoary bat, an endemic subspecies of a wide-ranging species found in temperate areas of North and South America. A good disperser, the hoary bat has become established in several island groups. One subspecies occurs in Ecuador's Galápagos archipelago.

The Hawaiian bat's ancestors arrived across more than 2000 miles of the Pacific Ocean. This bat was probably never very common and its population is now estimated at a few thousand animals. Because of its solitary and nocturnal habits and its low numbers, it is not often seen. Though less common on Maui than on Kaua'i and Hawai'i, bats do occur over a range of upper elevation areas in Haleakalā National Park. The Hawaiian bat is sighted on rare occasions along the rim of Haleakalā crater from Halemau'u trail to the summit, as well as in Kaupō Gap and Kīpahulu Valley.

Unlike many bats, the Hawaiian bat is solitary, not social. It is a small reddish-gray creature weighing just over half an ounce. The Hawaiian bat does not inhabit the dark interiors of caves or lava tubes, but roosts in trees or among rocks during daylight hours. It hangs upside down, grasping a branch or other support with its feet, and curling its furry tail over its head like an umbrella, presumably to shed rain. At night it uses high-pitched cries and sonar to locate the flying insects upon which it feeds. The bat's Hawaiian name, *'ōpe'ape'a*, may be derived from the Hawaiian word pe'a, meaning "cross- shaped" or "sail-shaped."

Bats are beneficial animals. In many parts of the world, their consumption of flying insects helps keep agricultural pest species in check. As a result, there were attempts nearly one hundred years ago to introduce foreign bat species to Hawai'i for the control of insects. Fortunately, these attempts failed; today, the dangers of importing foreign animals are better appreciated. Importation of foreign bats could bring rabies and other diseases to people as well as to native bats in Hawai'i.

Unlike many endemic animals, the *'ōpe'ape'a* has adapted somewhat to the changed, modern landscape of the Hawaiian Islands. The Hawaiian bat is not limited to native *'ōhi'a (Metrosideros polymorpha)* and *koa (Acacia koa)* forests, but is also found in middle elevation sites such as 'Ulupalakua, Kula, and Makawao. Bats are not very selective in choosing trees for roosting and even use planted macadamia and eucalyptus trees. The *'ōpe'ape'a* is one of the few native animals to survive in lowland areas in Hawai'i, even though its original habitat has been much changed by development and agriculture.

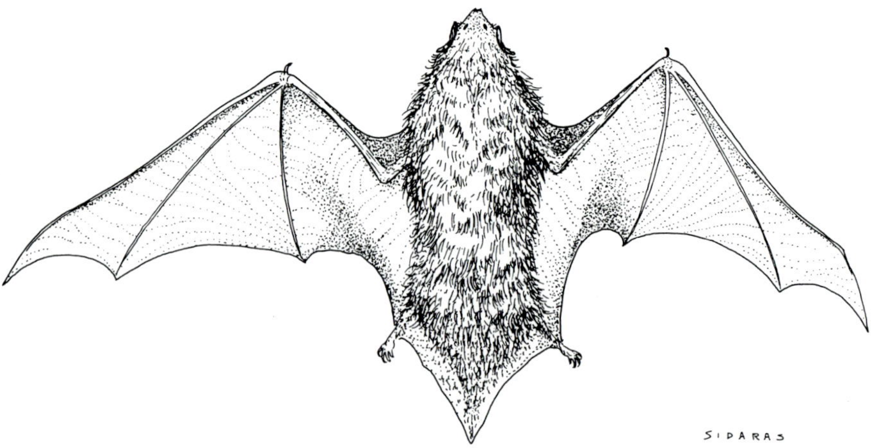

'ŌPE'APE'A
Hawaiian hoary bat
Lasiurus cinereus semotus
Common bat family (Vespertilionidae)

III. BIRDS

The birds of the Hawaiian Islands today are far different from those present in the islands before humans arrived. At Haleakalā National Park, more than half the bird species currently found within the park boundaries have been introduced by humans. Of these, the chukar partridge *(Alectoris chukar)* from Asia and the ring-necked pheasant *(Phasianus colchicus)*, native to China and Japan but widely introduced throughout the world, are two of the most common large birds in the park.

For many groups of Hawaiian plants and animals, there is only speculative information about their former diversity when the first Polynesian settlers arrived in approximately A.D. 400. With birds, there is the solid evidence of their semi-fossilized bones either buried in sand and soil or on the surface in lava tubes, sinkholes, and sand dunes. These new and very exciting discoveries are primarily the work of two scientists from the Smithsonian Institution, Washington, D.C. Their work has revealed many new species of Hawaiian birds. They have also discovered that other Hawaiian birds now very rare and restricted to isolated mountain cloud forests were once much more common, even down to the coast. Within the latest decade, we have come to more fully realize the magnitude of the declines and extinctions of Hawaiian birds.

Most of the sites where bird bones have been found are in lowland areas not occupied by native birds in modern times. The discovery that an abundant and diverse Hawaiian bird fauna formerly occupied lowland areas was unexpected. Some of the same bird species found in the fossil record still survive, but are now extremely rare and restricted to small populations in the most pristine and remote upland rain forests of Hawai'i. Some birds, such as the vermilion sickle-billed *'i'iwi*, were observed visiting the flowers of coconut trees in Kona as late as the 1890s, but have since disappeared from the lowlands entirely. As the bones collected by the Smithsonian research team were sorted and identified, it became clear that there once existed many Hawaiian birds that were never seen by Westerners. These endemic birds were already extinct by the time Captain Cook's ships arrived in 1778. These extinct Hawaiian birds not recorded historically included nine species of ducks and geese, ten species of flightless rails, two species of large crows, four species of long-legged owls, three species of flightless ibis, eighteen species of honeycreepers, and one species each of honeyeater, hawk, and eagle.

Some birds that once inhabited the park are known only through these fossil records. The crows, thrushes, geese, seabirds, and extinct honeyeaters and honeycreepers discovered in lava tubes on privately-owned ranch land on leeward Haleakalā likely occurred within park boundaries as well. In recent explorations of lava tubes in the rain forests of the park in Kīpahulu Valley, it was discovered that both the *nēnē* (p. 13) and the *'ua'u* (p. 12) formerly occurred in the rain forests as well as in the upper-elevation lava and cinder areas of the park. Also found in these park lava tubes were the bones of a new species of extinct Haleakalā honeycreeper *(Orthiospiza howarthi)*, a robust thick-billed finch of the type last seen in Kona at the turn of the century, which fed primarily on the seeds of *koa* and other trees. In the lava tubes also were the skeletons of flightless ibis (*Apteribis brevis*), flightless rail *(Porzana)*, and flightless goose *(Ptaiochen pau)*. These birds, previously known from remains found elsewhere on Maui, are new records for the park. The birds had vanished from even this remote area of pristine rain forest without documentation of their existence by Europeans.

What caused the loss of so many Hawaiian birds? Polynesian settlers used the lowlands for agriculture and in so doing gradually destroyed the lowland Hawaiian forest habitats of birds, especially on the leeward sides of the islands. In addition, native birds (especially the flightless forms) were used as food, a commodity increasingly in short supply as the human population of the Hawaiian Islands reached its peak before western contact in 1778. The exact population of Hawaiians before 1778 is uncertain (two hundred thousand to one million), but in the late eighteenth century, Hawai'i clearly had the largest human population in Polynesia.

With the arrival of Europeans came new plants and animals — and a new set of problems for the native biota. Since then, the extinction of native Hawaiian birds has continued virtually unabated. Recently, land management efforts have been initiated to protect areas of intact *koa* and rain forests and their many associated native plant and animal species.

The evolution of many species from a single ancestor, termed "adaptive radiation" by evolutionary biologists, occurs most strikingly on islands. The development of the Hawaiian bird group now called the drepanids or honeycreepers is similar to, but much more complex than, the evolution of the much-publicized Galápagos finches. Of the estimated 52 species of Hawaiian honeycreepers, 18 species (35%) are known only from the fossil record and were probably extinct by 1778. Twelve species (23%) have been eliminated since 1778. Of the 22 surviving species, 14 are classified by the U. S. Fish and Wildlife Service as endangered species; several of these may already be extinct.

Today, eight species of honeycreepers occur within Haleakalā National Park, most of them in the rain forests of the eastern and northeastern slopes; half of these are endangered species. The most common honeycreepers on Maui and in the State are the crimson-red, black and white *'apapane* (5 inches long) *(Himatione sanguinea)* and the yellow-green *'amakihi* (4.5 inches long) *(Hemignathus virens)*. Their ranges extend from the rain forests into upper-elevation shrublands and even into planted forests. The *'i'iwi* (6 inches long) *(Vestiaria coccinea)* is less common but still widespread; with its strongly curved bill, scarlet-orange plumage, and piercing whistle-calls, it is one of the most beautiful of the honeycreepers. All three birds, especially the *'apapane*, may be seen along the highway near headquarters and above to 9000 feet elevation, often flying across the road in search of the clusters of bright yellow flowers of the *māmane* tree *(Sophora chrysophylla)*. Although these three birds do eat insects and spiders, they are strongly nectivorous, drinking nectar that they siphon from flowers with their long, infolded "drinking straw-like" tongues.

The rain forests of upper Haleakalā and the Alaka'i swamp on Kaua'i are the best places in the islands to see the Hawaiian creepers, also in the unique honeycreeper family. The females of the small, plump Maui creeper

(Paroreomyza montana) (4.5 inches long), or *'alauahio*, are green; the males yellow-green. These birds forage in pairs or small flocks mostly in the sub-canopies and understories of trees and shrubs, making a characteristic "chip" call and feeding on insects, spiders, and less often, nectar of *'ōhelo (Vaccinium calycinum)* and *'ōhi'a* flowers. Occasionally, a Maui creeper may be seen 60 feet above the forest floor, gleaning the pale bark of a tall *koa* tree.

The *'ākepa (Loxops coccineus)* honeycreeper has dull gray-green females. Males change from green to progressively brighter, more entirely orange plumage with age. The *'ākepa* bill is short and crossed at the tip, enabling the bird to forage for caterpillars and other insects and spiders in the leaf tips in the canopies of *'ōhi'a* and *koa* trees. In 1894, the naturalist R.C.L. Perkins considered this bird "abundant" in the rain forests below Hosmer Grove at 5000 feet elevation, an area from which it has now apparently disappeared. Today, the *'ākepa* (which means "sprightly") is a federally recognized endangered species; only about two hundred of these small birds (4 inches long) were estimated to remain in the rain forests of northern Haleakalā in 1980. However, no *'ākepa* were seen in an intensive survey in 1992.

In 1973, members of the Upper Hāna Rain Forest research project, in exploring the remote northern and northeastern forested slopes of upper Haleakalā, noticed a few unidentified, small (5.5 inches long), black, brown, and white tri-colored birds creeping secretively through dense foliage of rain forest shrubbery. No new Hawaiian bird had been described for many years, but this discovery in 1973 of a bird never before seen led to the description of a new genus and the latest honeycreeper species, *Melamprosops phaeosoma*. Hawaiian language expert Mary Kawena Pukui named the new bird *po'o-uli*, literally "dark-headed." Based on recent finds of its bones in lava tubes, there is evidence to suggest that the *po'o-uli* was apparently once much more common and widely distributed on Maui. Sadly and ironically, this newly found species may soon be lost again, but this time forever. On lands adjoining the park, the *po'o-uli*, an endangered species, teeters at the edge of extinction with the entire population estimated to consist of only 140 birds in 1980. There have been few recent sightings, and numbers are believed to have declined drastically since 1980. This species appears precariously close to extinction.

Within Haleakalā National Park and its immediate environs, the populations of three other endangered bird species — the crested honeycreeper (p. 7), Maui parrotbill (p. 9), and the Maui *nukupu'u* (p. 9) — contain the last survivors of their kinds.

'ĀKOHEKOHE

Crested honeycreeper
Palmeria dolei
Cardueline finch family (Fringillidae)
Hawaiian honeycreeper subfamily (Drepanidinae)

The crested honeycreeper, *'ākohekohe* was once found on both Maui and Moloka'i but is now limited to East Maui. It is found in pristine high-elevation rain forests (4200-7100 feet elevation) of Haleakalā, but it is rare. The U. S. Fish and Wildlife Service estimated its population at 3800 birds in 1980.

The distinctive head feathers (light gray to orange) of this comparatively large forest bird (6.5-7.0 inches long) are tufted forward, forming the characteristic crest for which the bird is named. The predominantly black body feathers are tipped with orange, gray, and white, giving it a scaled appearance. These are bold, aggressive birds that will

ʻĀKOHEKOHE
Crested honeycreeper
Palmeria dolei
Cardueline finch family (Fringillidae)
Hawaiian honeycreeper subfamily (Drepanidinae)

drive away other honeycreepers, especially *ʻapapane* and *ʻiʻiwi*, from flowering plants within their territories. This species and other honeycreepers often feed on nectar from flowers of the *ʻōhiʻa*, the most abundant tree in Hawaiian forests. As with many honeycreepers, the tongue is long and infolded to form a siphon-like tube for drinking nectar from flowers. Like many Hawaiian honeycreepers, *ʻākohekohe* also eat insects such as caterpillars.

Disease is one of many factors blamed in the decline of the *ʻākohekohe*. Like many Hawaiian honeycreepers, the crested honeycreeper is thought to be highly susceptible to avian malaria, a disease introduced early in this century with alien birds. The disease is transmitted by an alien mosquito and is believed to prevent honeycreeper populations from surviving in lower elevation forests. Mosquitoes rarely occur above 4000 feet elevation; many native birds escape infection only while they remain in upper elevation forests.

MAUI PARROTBILL

Pseudonestor xanthophrys

Cardueline finch family (Fringillidae)

Hawaiian honeycreeper subfamily (Drepanidinae)

The Maui parrotbill is one of the most unusual results of honeycreeper evolution. This very rare bird, endemic to Haleakalā, is found in understory shrubs and trees of windward *'ōhi'a* and *'ōhi'a-koa* rain forests mostly between 4000 and 6800 feet elevation. This small, predominantly olive-green bird (5.5 inches long) with yellow and brown markings is characterized by a short tail, bold and dark eye stripe, and large parrot-like bill. The species was already rare by the 1890s in *koa*-dominated forests at 4000-5000 feet. When these *koa* forests were destroyed, the parrotbill survived only in higher-elevation *'ōhi'a* forests.

The Maui parrotbill feeds largely on larvae and pupae of the wood-boring beetle *Plagithmysus* (p. 20) and other insect larvae that live in damaged wood of forest trees such as *koa, alani* (*Melicope* spp.), *'ākala (Rubus hawaiiensis)* and *'ōlapa (Cheirodendron trigynum)*. The Maui parrotbill is often found foraging in mixed flocks with the Maui creeper and formerly the Maui *nukupu'u*. While feeding, the bird is active and agile, parrot-like in both behavior and appearance. It searches up and down twigs and small branches, often in contorted positions, even swinging completely upside down. Compared to many of its honeycreeper relatives, the Maui parrotbill is a slow and methodical bird, described by R.C.L. Perkins as "unsuspicious." According to the naturalist, once the bird decides on a branch, it seizes it with its massive overextended bill and, using its oversized jaw musculature first grips, then splits or cracks the hard, dried wood, a feat described as requiring "enormous strength." This accomplished, the bird eats the pale beetle larvae that live in galleries within the wood. While collecting specimens for study in 1894, Perkins noted that the stomachs of Maui parrotbills were often "packed full" with *Plagithmysus* beetle larvae, each up to 1 inch long.

The U. S. Fish and Wildlife Service estimated the population of this endangered species in 1980 at five hundred birds in the *'ōhi'a* rain forest on the windward flank of Haleakalā. The Maui parrotbill is still found in fringes of *koa* forests in the Waikamoi Preserve below Hosmer Grove and in the rain forests of upper Hāna and Kīpahulu within the park. According to naturalist George Munro, this bird was reported in the 1930s to visit groups of trees in eastern Kaupō Gap from its home deep in the rain forests. It is the last of at least five species of hook-billed honeycreepers that were once found on Maui.

MAUI *NUKUPU'U*

Hemignathus lucidus

Cardueline finch family (Fringillidae)

Hawaiian honeycreeper subfamily (Drepanidinae)

The small, yellow and olive-green *nukupu'u* (4.5-5.5 inches long) was extirpated on O'ahu by 1860 and last seen on Kaua'i in 1986. In the 1890s, the *nukupu'u* was still frequently seen in koa forests of East Maui at 4000-4500 feet elevation. Today, the Maui *nukupu'u* is extremely rare or perhaps extinct; the last birds were seen in the 1980s.

The *nukupu'u* is similar to another endangered species, the *'akiapōlā'au* of the island of Hawai'i. That species uses its straight lower bill to hammer tree bark much like a woodpecker while searching for insects, making a tapping sound that early naturalists used in locating the birds. The Maui *nukupu'u* is less vigorous in probing with its lower bill and, Perkins noted, feeds almost silently. The bird uses its long, curved bill

MAUI PARROTBILL

Pseudonestor xanthophrys

Cardueline finch family (Fringillidae)

Hawaiian honeycreeper subfamily (Drepanidinae)

MAUI *NUKUPU'U*
Hemignathus lucidus
Cardueline finch family (Fringillidae)
Hawaiian honeycreeper subfamily (Drepanidinae)

to sweep newly opened crevices in the bark in search of spiders, caterpillars, *Oodemas* weevils (p. 22), and other insects.

Sightings of the *nukupu'u* have been very infrequent in recent years. It was estimated in 1980 based on limited sightings that less than thirty birds survive on upper Haleakalā; during the recent 1992 Maui bird survey, no *nukupu'u* were reported. *Koa* forests, the preferred habitat of *nukupu'u*, have declined drastically and now occupy only a small percentage of their original range (largely at higher elevations). Wildfires, browsing by introduced mammals, and *koa* logging were responsible for the loss of the great *koa* forests that once grew on the western flanks of Haleakalā. Sightings of this bird on Maui in the 1890s were in the extensive *koa* forests that still occurred above Makawao and below Hosmer Grove, near 5000 feet elevation. However, with the destruction of this forest and the presence of malaria-carrying mosquitoes at lower elevations, all recent sighting of these birds have been made above 5000 feet elevation in forests dominated by *'ōhi'a*.

'UA'U

Hawaiian dark-rumped petrel
Pterodroma phaeopygia sandwichensis
Petrel family (Procellariidae)

The dark-rumped petrel *(Pterodroma phaeopygia)* nests in only two areas in the world — the Hawaiian and the Galápagos archipelagoes. These seabirds were once extremely abundant on all the main Hawaiian Islands. In former times, such numbers of *'ua'u* returned nightly to their nests on the island of Moloka'i that the incoming birds were said to have "darkened the skies." The species either no longer occurs or else is extremely rare on Moloka'i and in most of its formerly occupied range. Predation by man and introduced mammals (primarily mongoose and cats) is probably the reason. Fewer than two thousand birds remain, almost all nesting in Haleakalā National Park. Breeding populations of unknown size nest at high elevations on Kaua'i and Hawai'i islands.

The *'ua'u* is rarely seen even by those most familiar with its habits. The bird's Hawaiian name, *'ua'u*, is derived from its haunting call. On warm spring and summer evenings, these seabirds call as they return to their nests from the sea and while soaring around the cliffs above the visitor cabins in Haleakalā. Campers at Hōlua in Haleakalā crater between March and September hear the eerie, resonant cries of this seabird starting about an hour after sunset and often continuing to 1:00 A.M.

'UA'U
Hawaiian dark-rumped petrel
Pterodroma phaeopygia sandwichensis
Petrel family (Procellariidae)

The *'ua'u* relies on precise timing and strong pair bonding between male and female birds to survive in its harsh world of volcanic cliffs and open ocean. The timing of events that accompany reproduction is critical; if something is askew, the birds will abandon nesting attempts for that year. The *'ua'u* lays a single egg each year, resulting in an extremely low reproductive rate. However, each *'ua'u* may live as long as twenty-five to thirty years.

The *'ua'u* arrive in Hawai'i beginning in February each year, returning from somewhere, perhaps the western Pacific Ocean, to nest high on the slopes of Haleakalā. The *'ua'u* nest in burrows which usually extend 3 to 6 feet, but may be up to 30 feet long. These burrows, dug with the bird's legs and wings, are usually in cinder, under rock outcrops, or under large boulders on faces of volcanic cliffs. About seven hundred established burrows are known in the park, used mostly by birds which have nested there year after year.

Egg-laying occurs in late April and early May. Male and female birds take turns incubating at the nest and feeding at sea. The foraging adult hunts at night, predominantly for squid, but also for certain deep-water fish that come to the surface at night. Incubation shifts usually last twelve to sixteen days, but periods as long as twenty-three days have been recorded. After the single egg hatches in about fifty-five days, both parents begin searching for food at sea, since great quantities of food are needed by the voracious chick. By early October, the chick, looking much like a plump ball of down, is sometimes twice the weight of an adult bird.

In late October, the parent birds leave and the chick takes on the distinctive black and white plumage of the adult *'ua'u*. After exploring the area around its burrow and exercising its wings, the young *'ua'u* takes its first flight. The fledgling bird leaves the burrow with other departing birds and heads for the open ocean. If all goes well, after about three to six years of maturing the bird finds a mate and begins the nesting process on the cliffs of Haleakalā.

Inexperienced birds making their first flights sometimes become disoriented and attracted to the bright lights of urban areas where they can become grounded and unable to fly. Over the years, many of these fledgling birds have been rescued by Maui residents and turned over to the Hawai'i Division of Forestry and Wildlife, where, if they survive, they are rehabilitated and released back to the wild.

In 1979, researchers found that predators (mongooses and cats) were destroying 40 percent of the yearly production of *'ua'u* eggs and chicks. Adult birds are also killed but often not eaten by mongooses. For a species with so few breeding pairs, and those few having a potential of producing only a single chick per year, predation can be catastrophic. A predator trapping program initiated in 1980 has now reduced predation on *'ua'u* to near zero. With this continued trapping effort by National Park Service personnel and volunteers, the long-term survival of this rare Hawaiian seabird seems probable.

NĒNĒ

Hawaiian goose
Branta sandvicensis
Goose family (Brantidae)

Haleakalā National Park is one of the few places in the world where it is possible to see individuals of the world's rarest goose living in its native habitat. At Haleakalā, *nēnē* can be seen especially around park headquarters and near cabins in the crater. There are only an estimated 200 to 250 *nēnē* on Maui, 350 birds on the island of Hawai'i, and

NĒNĒ
Hawaiian goose
Branta sandvicensis
Goose family (Brantidae)

about 50 wild *nēnē* on Kaua'i for a total population of only about 600 wild birds.

The closest living relative of the *nēnē* is probably the Canada goose, *Branta canadensis*. However, other geese and goose-like birds have existed in the Hawaiian Islands and have become extinct. On Maui, there were four other species besides the *nēnē*. Two species were smaller birds capable of flight, while two others were larger, flightless species. In the history of all the Hawaiian Islands, there were probably about a dozen species of geese and goose-like birds, many flightless or very poor fliers. Of these, only the *nēnē* survived into this century.

The *nēnē* is the state bird of Hawai'i and was once much more common than it is today. Recent discoveries of fossil bones have shown that *nēnē* were formerly found not only on Maui and Hawai'i, but also on Kaua'i, O'ahu, and Moloka'i.

The *nēnē* population statewide plummeted drastically in modern times. *Nēnē* were eliminated from Maui by the 1890s. In 1944 it was estimated that only thirty *nēnē* were left in the wild. In 1946 and 1947, wildlife biologists C.W. and E.R. Schwartz reported no *nēnē* sightings in eighteen months of field work on Hawai'i. They wrote, "This wildfowl is the next Hawaiian, if not world, species facing imminent extinction. To permit this tragedy to occur without exerting more effort than has been [done] to date is unpardonable."

By 1949 the number of *nēnē* was less than thirty; after completing their fieldwork, the Schwartzes recommended trying to save the species through intensive captive breeding. One breeding program was started in 1949 at Pōhakuloa on Hawai'i and another in 1950 at the Wildfowl Trust at Slimbridge in England. Though the Hawaiian territorial government had been raising *nēnē* since 1927, the program began in earnest with the inception of the Pōhakuloa program. From only six birds in 1949, the population at Pōhakuloa gradually grew and eventually produced nearly 1700 *nēnē* in the period from 1949 to 1978. Of that total, 285 *nēnē* were released in Haleakalā crater. The Wildfowl Trust at Slimbridge raised 782 birds over a twenty-eight year period; of these, 197 were reintroduced into the wild in Hawai'i, all at Palikū in Haleakalā National Park. In total, with a few geese added from a private collection, 489 *nēnē* were released at Haleakalā between 1962 and 1978. During the 1980s, captive populations have been maintained in England and Hawai'i, but few additional birds have been released into the wild on Maui.

The *nēnē* population within Haleakalā National Park now consists of only approximately 200 to 250 birds. Many released birds have died, but many wild birds are unbanded and were apparently born in the wild. On Hawai'i, 1271 *nēnē* were released between 1960 and 1978; less than one third of that number are alive today.

Nēnē evolved in a world without predators, except for owls, hawks, and eagles (the latter known in the islands only as fossils). In continental areas, ducks and geese avoid predation by escaping to large areas of water; by synchronous breeding; and by short courtship, nesting, and post-nesting periods. Breeding in areas without water was no problem for ancestors of today's *nēnē* because there were then no mammalian predators in Hawai'i. *Nēnē*, however, have a prolonged breeding period of three to six months (October through March) and the longest period of incubation (thirty days) and rearing of young (ten to twelve weeks to fledging) among the world's geese. In these respects, the *nēnē* is now poorly adapted to the mammalian predators that have been introduced by humans to Hawaiian habitats.

Observations made prior to the 1900s indicated that *nēnē* bred primarily in the lowlands, then moved fledglings to higher elevations. The loss of predator-free lowland habitat in Hawai'i, especially on Maui, has forced *nēnē* to keep to upper elevations alone. *Nēnē* nesting season is the wettest time of year in Hawai'i. In the leeward lowlands, the rains trigger a surge in plant growth, which would provide abundant food for *nēnē*. At upper elevations, however, the increase in plant food is marginal, and long periods of cold rains and strong winds make the rearing of young goslings difficult.

Like most geese, *nēnē* are grass grazers, eating both native and alien species. At Haleakalā, they also feed on the alien gosmore *(Hypochoeris radicata)* and less often on the fruits of three native shrubs: *ʻōhelo (Vaccinium reticulatum)*, *pūkiawe (Styphelia tameiameiae)*, and *pilo (Coprosma montana)*.

In spite of remarkable success in captive rearing of *nēnē* from 1950 to 1980, the future of this bird is uncertain because of poor reproductive success in the wild. The most promising strategies for the future involve continued research and localized control of predators. The removal of feral goats in Haleakalā National Park has led to increased vegetation in many areas; this increase of food may significantly improve lower elevation *nēnē* habitat, particularly in the Kaupō Gap area (4000-6000 feet elevation). Continuation of captive propagation and monitoring of *nēnē* in the wild, as well as close cooperation among state and federal agencies involved in *nēnē* protection, are essential if the Hawaiʻi state bird is to survive in the wild.

IV. INSECTS AND SPIDERS

The native Hawaiian insects and spiders are widely known among biologists for their remarkable adaptations and speciation. Prior to human contact, 99 percent of Hawaiian insects were found nowhere else on earth; they were endemic to the islands. As with other forms of native Hawaiian life, the Hawaiian insect fauna developed from few ancestors, largely due to the difficulties of long-distance dispersal across a formidable ocean barrier. Some groups of insects, including some very successful on the continents, failed to reach Hawaiʻi before the arrival of humans. As a result, there were no mosquitoes, ants, termites, honeybees, yellowjackets, fleas, or mayflies in the islands.

Once an insect successfully arrived in Hawaiʻi, it was freed from predators and competitors of its original habitat. Conditions in this new island home often led to rapid speciation. For example, the arrival of two individuals of vinegar fly *(Drosophila)* on Hawaiian shores millions of years ago may have been the events that sparked the evolution of approximately eight hundred and sixty endemic species.

Many of the insects that arrived in the Hawaiian Islands evolved features completely unknown in their continental relatives. An inability to fly, a special case of unusual diets, and other unique adaptations appeared in the Hawaiian species.

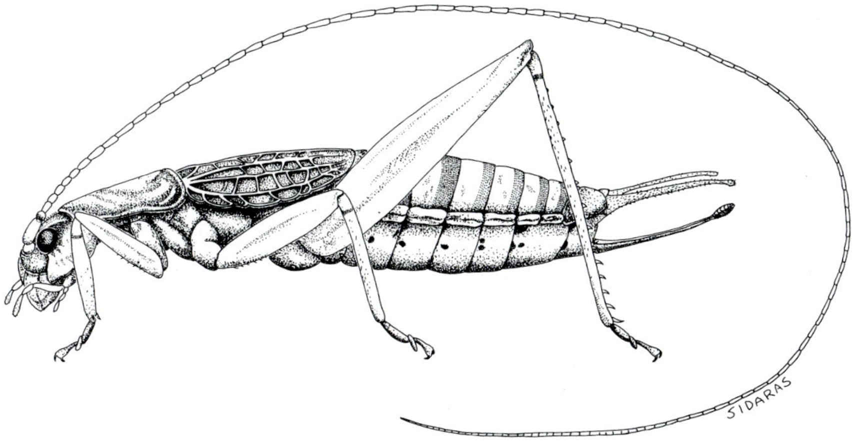

HAWAIIAN CRICKET
Prognathogryllus kipahulu
Order Orthoptera
family Gryllidae

HAWAIIAN KATYDIDS AND CRICKETS

Prognathogryllus kipahulu illustrated
Order Orthoptera, families
Tettigoniidae and Gryllidae

Among the world's representatives of this order, Hawaiian crickets and katydids are extremely diverse. Experts believe there may be several hundred native species in Hawai'i, mostly undescribed, more than occur throughout the remainder of the United States. In crickets, the male sings by rubbing his wings together to attract females of his own species. Each of the many Hawaiian cricket species has its own unique song by which members recognize each other. In the dark rain forest understory, where the ground is covered with *hō'i'o* fern *(Diplazium sandwichianum)* and every tree is covered with a thick coat of mosses, many cricket species combine their individual songs into a complex symphony of sound. Hawaiian crickets and katydids occur most abundantly in the *koa* and *'ōhi'a* rain forests from just above sea level to 5000 feet elevation.

Within the rain forests of the park occurs the endemic katydid *Banza mauiense*, 1.5 inches long and camouflaged a perfect leaf green. The Hawaiian crickets include the smaller, indigenous genus *Paratrigonidium* (1/2- to 3/4-inch long) and three larger endemic genera *Leptogryllus, Thaumatogryllus*, and *Prognathogryllus* (about 1-1.5 inches long). All Hawaiian crickets are brown and flightless and some are loud, strong "singers". In Hawai'i, there are two endemic genera of katydids and four genera of crickets, three of which are endemic.

Within Haleakalā National Park, most native katydids and crickets are found in the rain forests of the middle and upper elevations of Kīpahulu Valley, a closed scientific reserve. Even in this remote, isolated wilderness, many species are not fully protected. Recent research has shown that alien rodents (rats and mice) prey on these rare insects.

HALEAKALĀ FLIGHTLESS LACEWINGS

Pseudopsectra lobipennis illustrated
Order Neuroptera, family Hemerobiidae

Many insect and bird species that have evolved in isolation on islands either have reduced flight ability or have lost the power of flight altogether. In the Hawaiian Islands, flightlessness has evolved in over a dozen insect families and three bird families. Biologists suggest that species may gradually lose the ability to fly in island habitats where flying drains energy and, without mammalian predators, may be of little adaptive value. In stable habitats on isolated islands, flightlessness may have been an advantage, and many species have adapted accordingly.

Among the flightless insects that have evolved on Haleakalā, the brown lacewings are among the most spectacular. Three species of flightless lacewings are unique to Haleakalā's upper slopes. *Pseudopsectra lobipennis* has small, strap-like rear wings, and textured, thickened front wings that cup and protect the insect's body. In spite of alien rodent predators this rare insect still survives in the dense rain forests of Kīpahulu Valley within the park.

Pseudopsectra cookeorum is known from a single collection in 1945 near the summit of Haleakalā on the silversword relative, *kūpaoa (Dubautia menziesii)*. The wings of this extraordinary species, no longer used for flying, are covered with long spines that give it roughly the appearance of a miniature hedgehog. These unique bristling wings provide protection and perhaps by breaking up its silhouette camouflage the insect's body. The spiny lacewing has not been collected since 1945, and its status remains uncertain. It may be extinct.

The third species, *Nesothauma haleakalae*, lacks rear wings altogether and has front wings that are thick and shield-like. This species was last collected above Makawao in the 1870s and may also be extinct.

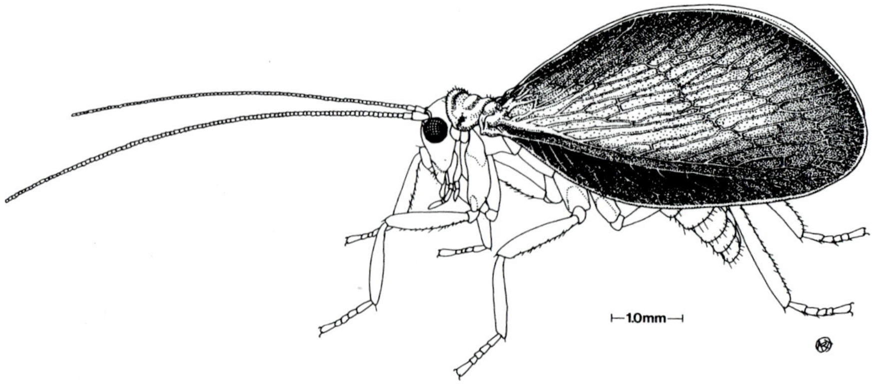

HALEAKALĀ FLIGHTLESS LACEWING
Pseudopsectra lobipennis
Order Neuroptera, family Hemerobiidae

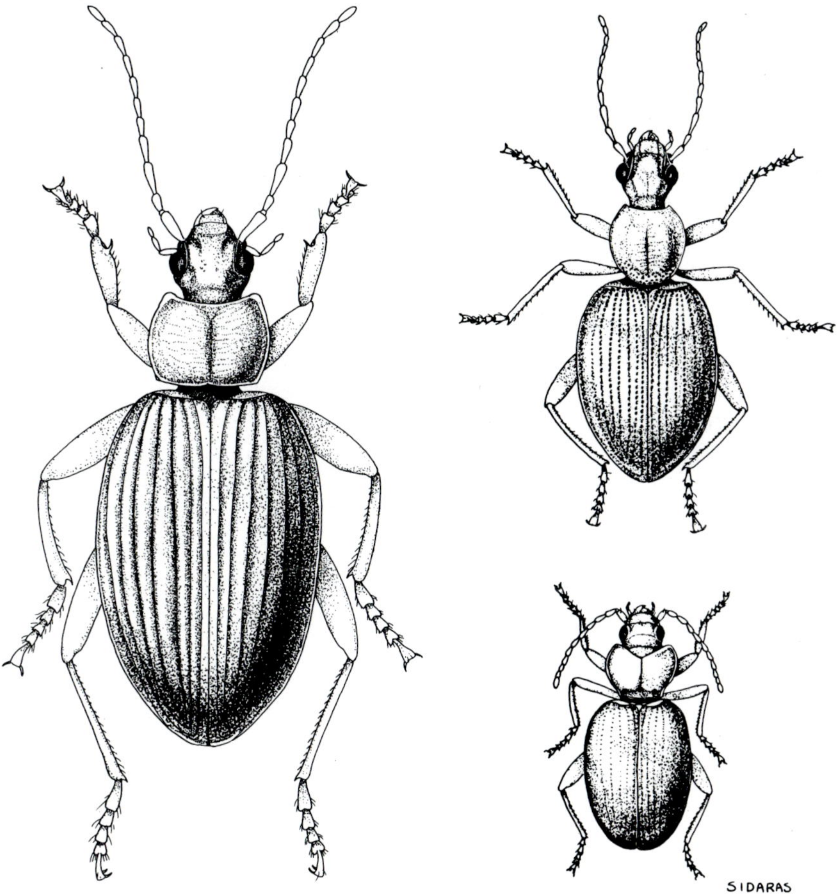

HAWAIIAN GROUND BEETLES
Barypristus rupicola (left), *Mecyclothorax robustus* (lower right), and *Mauna frigida* (upper right)
Order Coleoptera, family Carabidae

HAWAIIAN GROUND BEETLES
Barypristus rupicola, Mecyclothorax robustus, and *Mauna frigida* illustrated
Order Coleoptera, family Carabidae

Worldwide, carabids are a large important family of beetles, predacious on arthropods and snails. The ground beetles (Carabidae) are one of the more extensively developed insect families in the Hawaiian fauna. The 215 endemic species in the Hawaiian Islands are believed to have evolved from as few as six original immigrants. Sixty-eight endemic species in eleven genera are found in and near Haleakalā National Park. Three relatively common species in the high-elevation shrubland are illustrated here. *Mauna frigida* is the sole species in a genus endemic to Hawai'i; this shiny 1/2-inch long beetle occurs only on the upper slopes of Haleakalā. *Barypristus rupicola* is one of the largest of the Hawaiian carabids. This species is 3/4-inch long, brown-black, and a high-elevation specialist found in barren Haleakalā rockscapes. *Mecyclothorax robustus*, a member of a widespread genus, is a species endemic to upper Haleakalā. Though collected only a few times since its discovery in the 1860s, this beetle can still be found in the area above Hosmer Grove.

In the late 1980s, in the deep rain forests of Kīpahulu Valley, two deep cave entrances were discovered covered with vines, matted ferns, and other vegetation leading to two sizable lava tube caves. Besides fossil bones, scientists from the Bishop Museum in Honolulu working with colleagues at Cornell University soon discovered and described two new species of ground beetles, each restricted to the dark world of its lava tube and unknown until recently.

HAWAIIAN LONG-HORNED BEETLES

Plagithmysus terryi and
Plagithmysus cheirodendri illustrated
Order Coleoptera, family Cerambycidae

Long-horned beetles of the endemic genus *Plagithmysus* provide one of the more remarkable examples of adaptive radiation in Hawai'i. Over 136 species in the genus are believed to have evolved from a single ancestral species which immigrated millions of years ago probably from southwestern North America; thirty species have been recorded in and near the park. Larvae of this group are wood borers that feed within living, often damaged, trees. The adaptive radiation of the wood-boring long-horned beetles in the genus *Plagithmysus* has specialized them to such a degree that in most cases one beetle species has only one tree species as its host. In the Hawaiian Islands, the native long-horned beetles have adapted to twenty-seven different families of trees. One of the Hawaiian honeycreepers, the Maui parrotbill (p. 9), uses its bill to tear apart plant stems in search of the pale larvae of these beetles, one of its primary foods. The naturalist R. C. L. Perkins found that the stomachs of the parrotbills he collected in 1894 were filled with long-horned beetle larvae.

The most common endemic long-horned beetle on Haleakalā is *Plagithmysus funebris*, a species whose larvae are found only on the common native leguminous tree, *māmane (Sophora chrysophylla)*. This cricket-like beetle is black with silver linear markings and is most frequently seen on warm summer days. Adults sometimes fall prey to alien yellowjacket wasps.

Plagithmysus cheirodendri, which lives only in the forests of East Maui, feeds exclusively on the wood of *kōlea* trees *(Myrsine lessertiana)*. The first collection of this species was an adult beetle found on an *'ōlapa (Cheirodendron)* tree near Palikū cabin at the eastern end of Haleakalā crater, hence its Latin name. Since then, however, the species has been reared from larvae found in the wood of the *kōlea* tree, and the *kōlea* has been identified as its true host. Adult beetles are brightly patterned with brick-red and yellow markings. The females lay their long, spindle-shaped eggs in the bark of *kōlea* trees in the rain forests. On hatching, the larvae burrow into the wood, feeding and developing for a year or more. If they survive without being discovered by forest birds or parasites, these larvae eventually pupate, become adult beetles, and begin their search for mates and the correct host tree for laying eggs.

Although most long-horned beetles are found in trees, one species found only on upper Haleakalā has adapted to the famous silversword plant rather than to a tree as its host. The larvae of this beetle, *Plagithmysus terryi*, bore

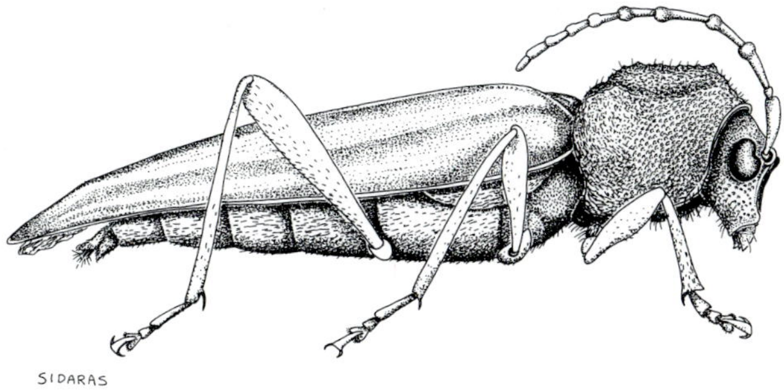

HAWAIIAN LONG-HORNED BEETLES
Plagithmysus terryi (above) and *Plagithmysus cheirodendri*
Order Coleoptera, family Cerambycidae

into the thick stem and root of both the Haleakalā silversword (p. 34) and the greensword (p. 36). Occasionally, silversword plants become so weakened by the boring that they topple, but usually no apparent damage is done. The beetle larvae live and develop inside the silversword plants feeding on the woody tissue until they pupate and then become adult beetles. The adult beetles, a little over an inch long, are dull black-brown to bright brick-red with no markings. They emerge to mate and lay eggs in the early summer months (from May to June) when winter rains diminish, days become warmer, and silverswords begin to flower.

HALEAKALĀ WEEVIL
Oodemas mauiense
Order Coleoptera, family Curculionidae

Snout beetles or weevils (Curculionidae) are plant-feeding beetles with long snouts; many are serious plant pests. The 58 species of small, rare *Oodemas* weevils of native shrublands and forests are found only in the Hawaiian Islands and seem to have no close relatives in the rest of the world. A favored food for birds, adult *Oodemas* beetles hide by day under bark and in mosses and leaf litter; at night, these tiny, shiny black rounded beetles (1/4-inch long) emerge to feed on native plants and to mate. At least 15 species of

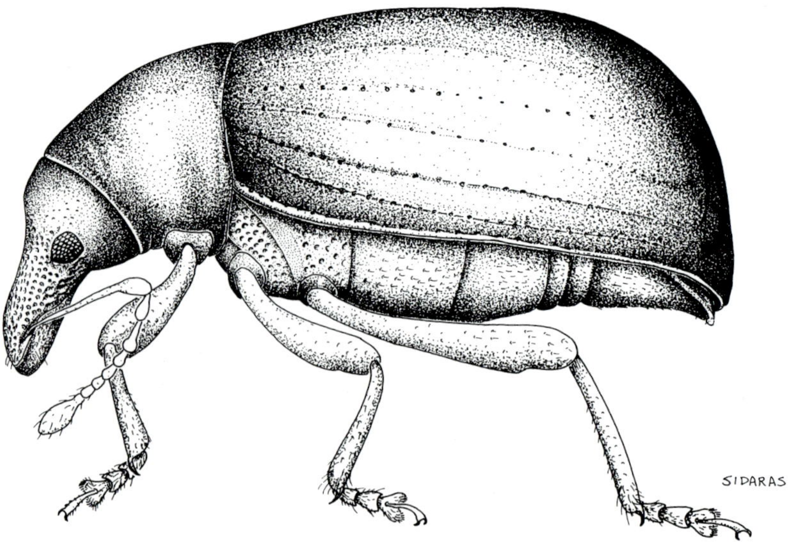

HALEAKALĀ WEEVIL
Oodemas mauiense
Order Coleoptera, family Curculionidae

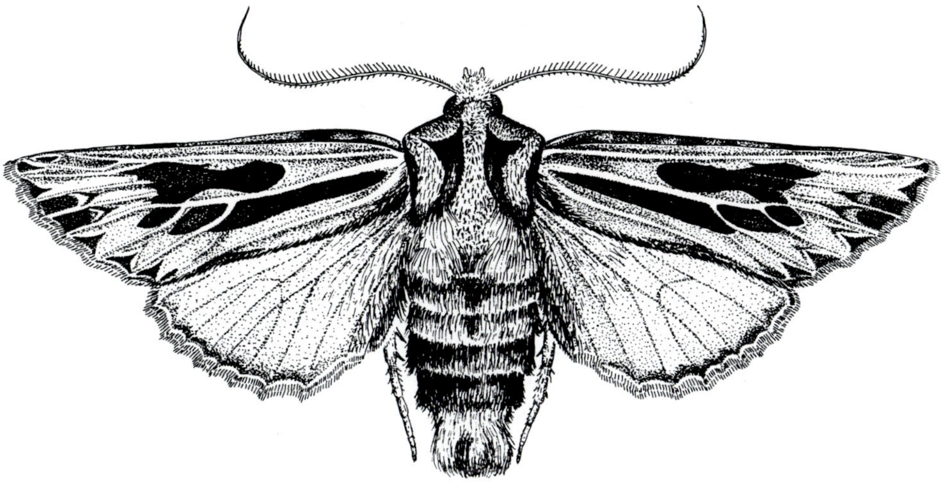

HAWAIIAN NOCTUID MOTH
Agrotis arenivolans
Order Lepidoptera, family Noctuidae

Oodemas are known from Haleakalā, either in the deep rain forests or in native shrublands from near headquarters to the mountain's summit.

HAWAIIAN NOCTUID MOTH

Agrotis spp.(*A. arenivolans* illustrated)
Order Lepidoptera, family Noctuidae

In their caterpillar form, noctuid moths constitute a large portion of the animal biomass produced on upper Haleakalā. They burrow in cinders during the day and at night feed on the common native shrubs *pūkiawe* and *kūpaoa*, the latter a silversword relative. Wherever the Argentine ant, an alien predator, is established, populations of caterpillars are nearly absent. Outside the ant's range, however, the 2-inch long native caterpillars are common. On warm, still summer nights, adult noctuid moths visit flowers, probably acting as pollinators for native plants.

HAWAIIAN CARNIVOROUS INCHWORM

Eupithecia scoriodes illustrated
Order Lepidoptera, family Geometridae

Throughout the world, the caterpillars of moths and butterflies, almost without exception, feed on plants. In the Hawaiian Islands, however, some species of moth larvae have abandoned this vegetarian diet and developed into carnivorous sit-and-wait predators.

In Haleakalā National Park, some species of the inchworm genus *Eupithecia* (such as *Eupithecia stypheliae)* feed on flowers and seeds. Others, including Haleakalā's *Eupithecia scoriodes*, have evolved a life style involving "ambush predation." These caterpillars,

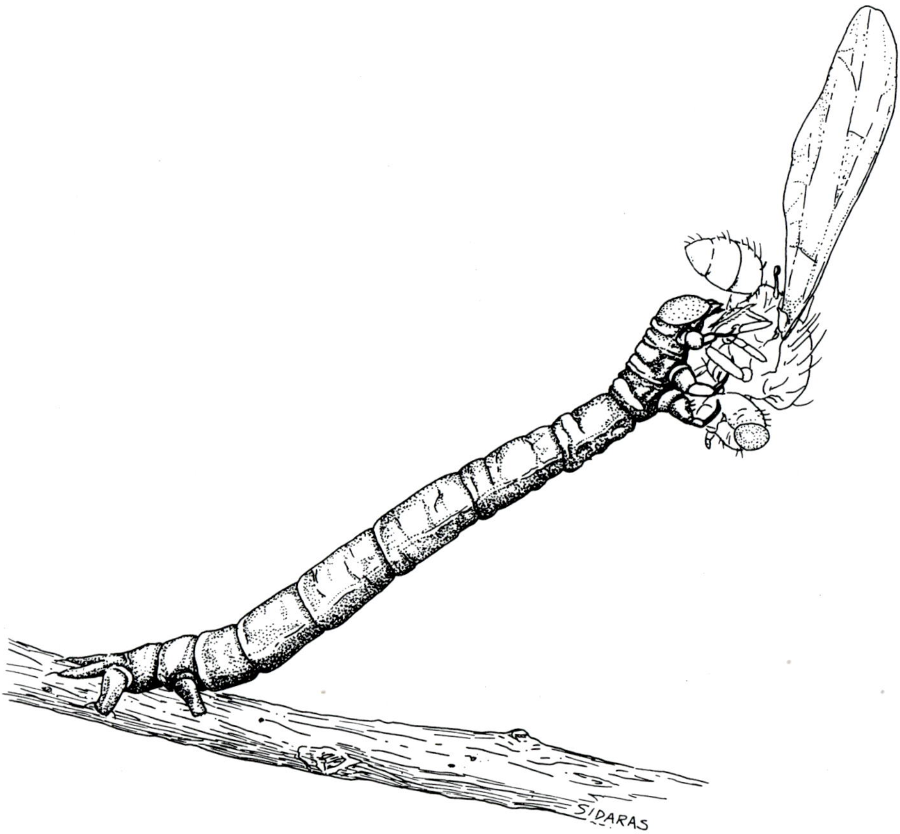

HAWAIIAN CARNIVOROUS INCHWORM
Eupithecia scoriodes
Order Lepidoptera, family Geometridae

usually about a half-inch long, settle on the edges of leaves or on plant stems to await their prey, which can be almost any tiny insect or spider that wanders along. While waiting, their bodies are held so stiffly as to make the small dark caterpillars appear very much like tiny twigs. When the unsuspecting prey touches the sensitive hairs at the tail end, the caterpillar instantly loops its head back, striking out with its large, claw-like legs to seize the prey. The attack is lightning-quick — as short as one-twelfth of a second. These caterpillars, so different in their diet and their feeding method from any other moth larvae in the world, eventually develop into attractive but quite ordinary looking inchworm moths.

HALEAKALĀ FLIGHTLESS MOTH
Hodegia apatela illustrated
Order Lepidoptera, family Gelechidae

Found only on upper Haleakalā at 8,000 to 10,000 feet elevation, this moth has narrow, dagger-like wings too small for flight. This unique species is believed to have evolved from a fully flighted species found on Haleakalā's lower slopes. It is such a rare and inconspicuous insect that after a single female was collected in the early 1900s, it was not re-collected until the mid-1970s.

The Haleakalā flightless moth is one of the rarest insects in the world; it is restricted to a few square miles of rocky, windswept habitat. Instead of flying, this silver-gray moth (about 1/2 inch long) walks and hops along the ground. With its small wings folded over its back, it superficially looks like a cricket. Haleakalā's flightless moths are most often seen on warm days in rocky areas, but are also attracted to lights at night. The hairy larvae are quite active and live in silken tubes woven in with dried *kūpaoa* leaves and small pieces of cinder. The flightless moth has been found among the prey of the alien western yellowjacket wasp *(Vespula pensylvanica)* and is likely also threatened by the alien Argentine ant *(Iridomyrmex humilis)*.

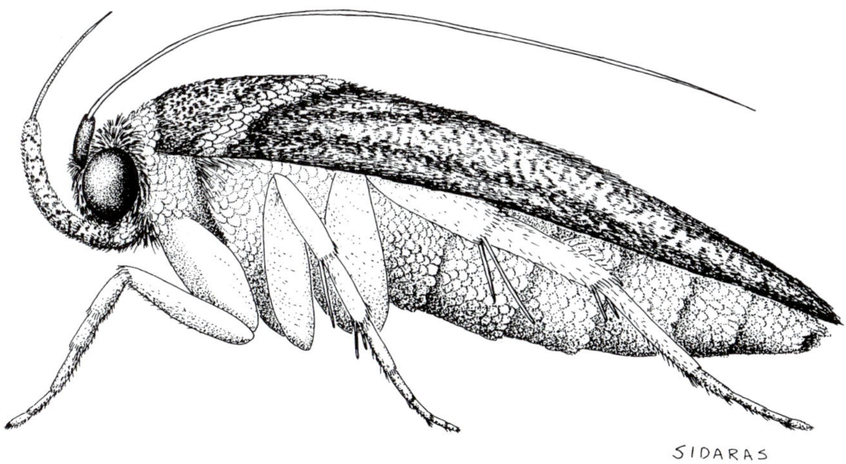

HALEAKALĀ FLIGHTLESS MOTH
Hodegia apatela
Order Lepidoptera, family Gelechidae

KAMEHAMEHA BUTTERFLY
Vanessa tameamea
Order Lepidoptera, family Nymphalidae

HAWAIIAN BUTTERFLIES
Vanessa tameamea illustrated and
Vaga (Udara) blackburni
Order Lepidoptera, families
Nymphalidae and Lycaenidae

The strikingly marked orange, black, and white Kamehameha butterfly is the larger of only two native butterflies of the Hawaiian Islands. The Kamehameha butterfly, or *pulelehua*, measures 2 inches across. The other native, the Blackburn's butterfly, is dark blue with bright-green underwings, and measures about 1 inch across. Both can be seen on warm days in the shrublands of Haleakalā's western slope. Larvae of the Kamehameha butterfly feed on the foliage of the *māmaki* bush (*Pipturus* spp.) and on other native plants of the nettle family (Urticaceae). Blackburn's butterfly caterpillars feed on the leaves and flowers of *'a'ali'i* bushes *(Dodonaea viscosa)* and *koa* trees. Adults of both species feed on the flower nectar of many native plants and likely serve as pollinators.

That Hawai'i has only two native butterflies can perhaps be explained by the fact that most butterflies are fragile insects and, as such, are not well suited for being carried thousands of miles across open ocean by the jet stream or other means. As with many endemic plants and insects, ancestors of the Kamehameha butterfly probably came from the southwestern Pacific area. The nearest relative of the Kamehameha butterfly is probably a species found across Europe, Asia, and eastward to the Marianas Islands.

About a dozen other butterfly species have been introduced to Hawai'i by humans. Those sometimes seen near park headquarters at 7000 feet elevation on Haleakalā's western slope include the painted lady *(Vanessa cardui)*, the painted beauty *(Vanessa virginiensis)*, the monarch *(Danaus plexippus)*, the white cabbage butterfly *(Pieris rapae)*, and the bean butterfly *(Lampides boeticus)*.

Inner cliffs of Haleakalā Crater provide a secure habitat for the Hawaiian dark-rumped petrel or *'ua'u (Pterodroma phaeopygia sandwichensis — p. 12).* Once common throughout the Hawaiian Islands, this seabird now nests primarily in deep burrows dug in cinder along cliffs on the upper slopes of Haleakalā National Park.

The chick of the rarely seen *'ua'u* may be twice as heavy as its adult parents. Its large size provides the surplus energy source to help quickly transform the fluffy chick into a mobile adult ready for the yearly exodus of petrels to a destination still unknown, somewhere in the western Pacific.

A striking "family" of simultaneously flowering Haleakalā silverswords, the *'āhinahina (Argyroxiphium sandwicense* subsp. *macrocephalum).*

Though there are an estimated 50,000 silverswords growing in Haleakalā Crater, the long-term survival of the species may be threatened by the elimination of their essential insect pollinators by an introduced predator, the Argentine ant, originally from South America, but now established near park headquarters.

Each thick rounded flower head (1-2.5 inches across) of the Haleakalā silversword contains hundreds of slim disk florets and a surrounding ring of magenta-bannered ray florets on a long stalk, clasped by a silver leaf-like bract. The nectar and yellow pollen of this and other flowers are the primary foods of the small black native bees, (p. 31) *Nesoprosopis,* pollinators which visit flowering plants in great numbers.

The large rounded rosettes of *Lobelia gloria-montis* (p. 39), shown here in rain forest of the upper wall of Kīpahulu Valley, are sometimes 3 ft across. The ancestors of this species may have arrived from western South America or eastern Asia.

The huge flowering spikes of *Lobelia gloria-montis*, studded with 4-inch long flowers with creamy white corolla and red-purple staminal columns, make flowering plants of this species visible for miles on a clear day.

In the cloud forest of upper Kīpahulu Valley, soft carpets of mosses and liverworts cloak the trunk of every tree and shrub. It is in this high-elevation refuge that many rare native forest birds (including Maui parrotbill, crested honeycreeper, *nukupu'u*, etc.) escape the deadly threat of avian malaria, transmitted primarily by the night-biting mosquito *(Culex quinquefasciatus)*.

The magenta-streaked flowers of *Geranium hanaense* provide bright accents for their sombre, montane bog habitats. Less than 700 plants of this rare species are known in two small bogs on the northeast outer slopes of Haleakalā.

Flowering quartet of greenswords *(Argyroxiphium grayanum)*. From these forest-dwelling greenswords, the related, brilliant silversword of cinder hills in Haleakalā Crater may have evolved. Greenswords occur only in montane bogs, treeless openings in otherwise tangled *'ōhi'a* forest where the rainfall averages about an inch per day.

Nectar from flowers of *'ōhi'a,* the most common tree of Hawaiian rain forests, provides one of the primary foods of our native forest birds. The flowers, or *lehua* as they are known by Hawaiians, are to many a proud symbol of Hawai'i's natural heritage.

The *'ākohekohe* or crested honeycreeper *(Palmeria dolei)* is currently known only from the tangled rain forests of Haleakalā's north and eastern flanks. A loud vocalist, the *'ākohekohe* has a broad repertoire of whistles, short songs, and low squawks, some of which can be confused with the other native Hawaiian birds, such as the very different looking *'apapane.*

The happy-face spiders *(Theridion grallator)* are named for the wide variety of bright patterns on the abdomens, some resembling smiling faces. These beautiful living jewels inhabit a delicate rain forest world, spinning small silken webs on the undersides of leaves. Though visually stunning, these spiders are small enough (total length including legs is 1/2 to 3/4 of an inch) that one needs a magnifying glass to really appreciate their beauty, "face to face."

The flowers of *Clermontia tuberculata*, found only in the rain forest of upper Haleakalā volcano, are covered with purple spines and bumps.

In the cool late afternoon, the rain forest of Kîpahulu Valley just below 5000 ft elevation is alive with the multiple tones of singing crickets (p. 17), many of them endemic species found only in this steep-walled, mist-filled valley.

The giant Hawaiian dragonfly *(Anax strenuus)* is the largest Hawaiian insect and one of the largest dragonflies in the world (wingspan sometimes exceeding 5.5 inches). Its Hawaiian name is *pinao*, the same word used to describe a wide-finned fish species.

Majestic *koa (Acacia koa)* forests, with trees up to 100 ft tall, are an important part of our national heritage protected within Haleakalā National Park. *Koa* forests support a wide array of plants, birds, and insects.

The arched flowers of the *koli'i (Trematolobelia macrostachys)* are visited for their sweet nectar by native birds whose foreheads can become smeared white with the sticky pollen.

Rosettes of the large lobelia *Cyanea hamatiflora* can be 6 ft across on palm-like stems sometimes over 25 ft tall. The very dark purple, strongly curved flowers (4 inches long) occur in dense clusters beneath the leafy rosette.

Until recently, feral pigs were severely damaging native forests throughout Haleakalā National Park. After construction of fences, such as this one in middle Kīpahulu Valley, it has been possible to partially control feral pigs. The forests are recovering well and the forest understory is once again filled with a thick carpet of native ferns.

HAWAIIAN YELLOW-FACED BEES
Nesoprosopis volcanicus illustrated
Order Hymenoptera, family Colletidae

To those familiar only with bumblebees or honeybees, tiny, black, shiny *Nesoprosopis* bees hardly seem like bees at all. Nonetheless, they are true bees which gather pollen and nectar from flowers to feed their young. Among all animals, bees are the most important pollinators of both crop plants and wildland plants. Their coevolution with plants has produced the bright, showy flowers of many plant species.

At the arrival of the first Europeans in Hawai'i, *Nesoprosopis* bees were found from the mountains to the shores. According to the noted Hawaiian biologist R.C.L. Perkins in the late 1800s, they numbered with the native solitary wasps in the "millions." Since then the arrival of foreign ants and the destruction of many leeward forests have left the numbers and diversity of species of these native bees much reduced. The upper slopes of Haleakalā National Park are one of the few places in the Hawaiian Islands where these bees are still common. The Hawaiian bees were named *Nesoprosopis*, meaning "island *Prosopis*" (a type of bee) in 1899. Unlike the social honeybee, Hawaiian bees are solitary, each female laying eggs in her own winding, silken tube nest, usually under a rock. In the park, three species live on the rocky slopes of upper Haleakalā. Two (*N. volcanicus* and *N. nivalis*) make nests and visit flowers to supply the cells with pollen and nectar. The third species *(N. volatilis)* is a nest parasite that lays its eggs in the already provisioned nest of one of the other two species; it may visit flowers, but does so only to gather nectar for its own energy needs.

The small (1/4- to 1/2-inch long) *Nesoprosopis* are critical in the pollination of many types of native plants. Without them, plants such as the common *pūkiawe* and the rare

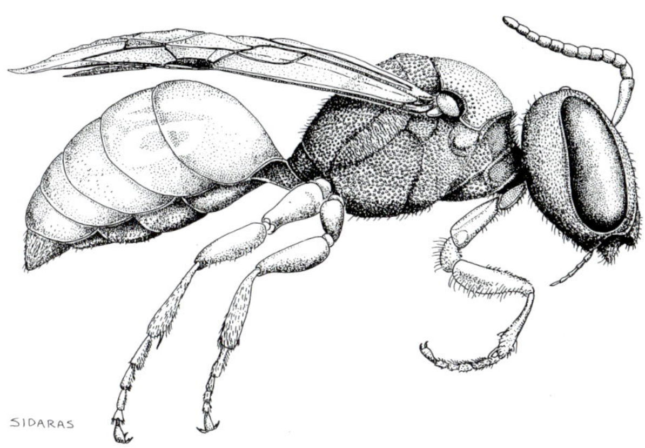

HAWAIIAN YELLOW-FACED BEE
Nesoprosopis volcanicus
Order Hymenoptera, family Colletidae

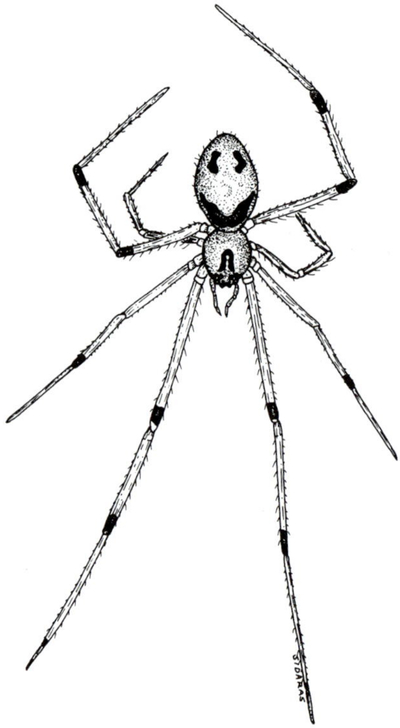

HAPPY-FACE SPIDER
Theridion grallator
Order Araneae, family Theridiidae

Haleakalā silversword would produce far fewer fertile seed. Within the park, the long-term survival of these insects is threatened by the recent (1960s) invasion of the Argentine ant, which is now common around park headquarters. Efforts to control the spread of the ant, which preys on underground nests of the native bees, may protect not only *Nesoprosopis,* but also the plants they pollinate.

HAPPY-FACE SPIDER
Theridion grallator
Order Araneae, family Theridiidae

The beautiful happy-face spider of Hawai'i is often depicted in books and magazines as a symbol of the state's natural heritage. Its common name refers to the wide variety of patterns, some resembling a smiling face, that occur on the top of the abdomen. These spiders and their "faces" are brightly colored with yellow, green, orange, and red markings. Recent research has suggested that the smiling-face patterns may be inherited characters. The color of a spider's body, however, can change every few days, depending on its diet.

Though found in many of the moist rain forests of Hawai'i, these spiders are quite uncommon and rarely seen. Living under the leaves of rain forest trees and shrubs such as *kanawao (Broussaisia arguta)*, *kāwa'u (Ilex anomala)*, and *'ōhāwai (Clermontia* spp.), these small (1/2- to 3/4-inch long) spiders spin long, silk threads that form an irregular mesh in which they catch their prey. The female spider, after mating with the slightly smaller male, retires to the underside of a leaf to lay an egg sac that contains ten to seventy eggs. The female vigorously defends first the eggs and then the tiny spiderlings, which are dependent on her for food and protection. To feed them, she lassos her prey with silken loops, leaving the ensnared victim (such as a small fly) for her young to eat.

V. PLANTS

The volcanoes that formed the present day Hawaiian Islands arose in one of the most isolated locations in the Pacific Ocean. Of the few plants that did successfully colonize those new islands, most (over 50 percent) arrived from the southwestern Pacific. However, in upper elevation areas such as Haleakalā where the weather is colder and more seasonal, many of the native plant ancestors originated in the Americas.

Just as with birds and insects, probably the most unique characteristic of Hawaiian plants is the adaptive radiation, or proliferation of species from one or few ancestors, that has occurred in certain groups such as the lobelias, tarweeds, and mints. Worldwide, perhaps the most dramatic example of adaptive radiation occurs among the Hawaiian tarweeds, known primarily by their most famous representative, the silversword. Consisting of 28 species in three genera, the Hawaiian tarweeds display incredible diversity of forms derived from a single ancestor which apparently arrived from California millions of years ago. These unique Hawaiian forms include the spectacular silversword — a gigantic, bright-silver, globe-shaped herb that flowers once, and dies; the related greensword; cushion plants; trailing and erect shrubs; rain forest trees; and even a vine species!

Another characteristic of Hawaiian plants is the importance of birds as pollinators, producing magnificent examples of coevolution. Today, many Hawaiian birds live primarily on the nectar of flowers of native plants, with a supplement of insects. The short finch bills of ancestral honeycreepers in some species have become long and curved, their tongues folded and tube-like. Indeed, the entire metabolism has adapted to the nectar diet. Likewise, many Hawaiian plants have adaptations to facilitate bird visitation, including production of abundant, protein-rich nectars; red flowers (which birds see most clearly); and curved, sickle-shaped corollas, such as those of the lobelias, mints, and geraniums.

Without any large native land mammals, Hawaiian plants evolved in the absence of pressures of browsing and grazing. As a result, another characteristic of Hawaiian vegetation is the lack of adaptations to deter browsing, such as large thorns, unpalatable textures, or noxious chemicals. With the introduction of cattle, goats, and pigs, many species of Hawaiian plants became rare or even extinct. One of the first steps needed to a protect native vegetation in a nature reserve is to exclude large browsing mammals that cause so much damage to native plants.

In total, there are about one thousand species of native Hawaiian flowering plants, about 90 percent of which are endemic. These species are believed to be derived from fewer than three hundred original ancestor species. There are also 170-180 species of native Hawaiian ferns; perhaps 70% of these are endemic. Haleakalā National Park contains about 250 species of native flowering plants and over 100 species of native ferns and fern allies.

ʻĀHINAHINA

Haleakalā silversword

Argyroxiphium sandwicense subsp. *macrocephalum*

Sunflower family (Asteraceae)

The Haleakalā silversword is a giant, silvery, rosette plant endemic to the crater and upper slopes of Haleakalā at elevations between 6,500 and 10,000 feet. The *ʻāhinahina* can mature from seedling to flower in as few as three years, but normally requires several decades, depending on the availability of moisture and nutrients in the soil. The plant remains a compact rosette until the final few months of its life, when it sends up an erect central flowering stalk, sets seed, and dies. The flower stalk, with one hundred to five hundred maroon-purple flower heads (each consisting of several hundred individual flowers), normally reaches a height of two to five feet, but occasionally grows to over eight feet tall.

This spectacular plant gained fame in the early years of Hawaiian tourism. Visitors to Haleakalā often took home entire plants to prove that they had reached the very summit. By 1920, silverswords had become so scarce on Haleakalā due to browsing by goats and cattle and human vandalism that the species was believed near extinction. Under protection since then, Haleakalā silverswords have thrived and increased. They now number about fifty thousand plants, mostly on cinder cones in Haleakalā crater. The silversword population and its environment are always changing, however; in some years, many silverswords become established, in other years, few or none. During particularly heavy rains, many seedlings are buried or washed away due to shifting cinders on steep slopes.

The Haleakalā silversword flowers in the warmer summer months of June through September. The number of plants that flower during any year has varied in the past 25 years from none to over 6600. Flying insects, especially native bugs, moths, flies, and bees, many of which are pollinators, are attracted in large numbers to the giant, aromatic inflorescence. About a dozen endemic species of moths, beetles, and other insects are found only with silverswords. All obtain their nourishment from live or decaying silverswords and some also act as pollinators. Botanists from the University of Hawaiʻi have recently shown that silverswords cannot produce fertile seed without cross-pollination. These rare plants, which flower only once after many years of growth, are dependent on insect pollinators for reproduction and hence long-term survival.

Despite a current population estimate of fifty thousand silverswords, the loss of pollinators caused by the invasion of a foreign ant could result in a dramatic reduction of seed set and the subsequent disappearance of natural populations of silverswords. The Argentine ant is an aggressive predator, recently established and slowly spreading on Haleakalā's western slope. Originally from South America, it preys on native insects of Haleakalā's high-elevation ecosystems, including important pollinators of the silversword. The Hawaiian Islands lack endemic ants, and native insects are not adapted and are therefore vulnerable to ant predation. This ant is presently found in only a small percentage of the upper elevation habitats of the park, but has the potential to spread throughout the crater and western slope.

ʻĀHINAHINA

Haleakalā silversword

Argyroxiphium sandwicense subsp. *macrocephalum*

Sunflower family (Asteraceae)

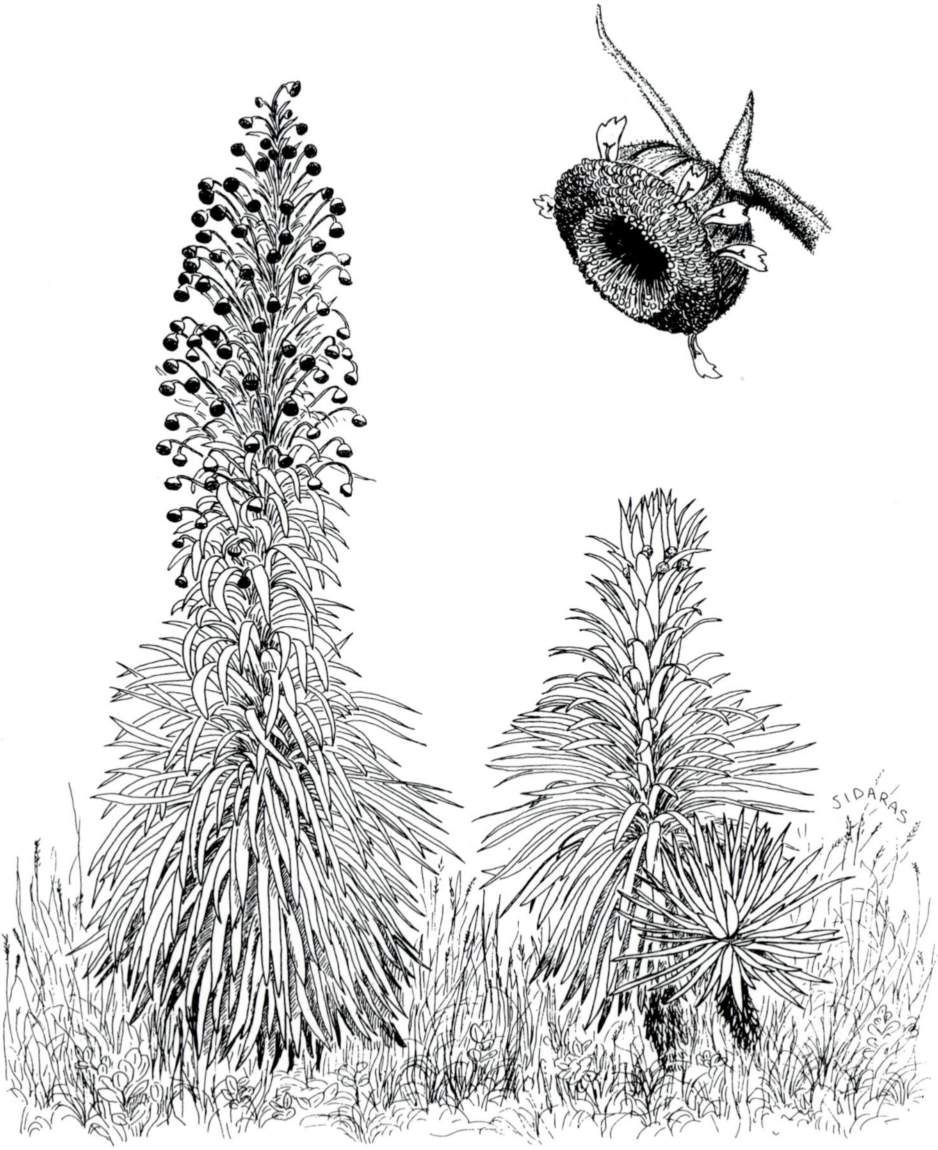

MAUI GREENSWORD

Argyroxiphium grayanum

Sunflower family (Asteraceae)

MAUI GREENSWORD

Argyroxiphium grayanum

Sunflower family (Asteraceae)

Found only on Maui, the greensword grows primarily in montane bogs, which are small treeless areas scattered through the upper-elevation rain forest at 5400-6600 feet elevation on northeastern Haleakalā, and at 4200-5700 feet elevation on West Maui. These areas are very wet with rainfall of 200-400 inches annually, so that ferns and bryophytes form thick carpets in the surrounding forests. These montane bogs are the habitats for unique native plants that grow nowhere else but in these small forest openings. Greenswords grow for ten to fifteen years; then, as a branch flowers, it dies. If as sometimes happens, a plant has only one large head (like the silversword) the entire plant dies. In areas protected from the wind, flowering plants can reach heights of over seven feet, but usually average three to five feet. The greensword flowers in the summer months of June through August; in some years, over a thousand greenswords may bloom, while in others, only very few.

Serious problems for greenswords in the montane bogs of the park began in the early 1970s. Feral pigs *(Sus scrofa)* from stock that escaped or were released from pig farms near the coast began to invade the high-elevation rain forests of upper Haleakalā. In their search for food (mostly certain plants, insects, earthworms, and mollusks), pigs extensively rooted many of the montane bogs of the park. In these rootings, aggressive foreign plants such as velvetgrass *(Holcus lanatus)* and gosmore began to grow and displace native plants. In bogs where feral pigs roamed freely, the cover of alien plant species rapidly increased. However, research demonstrated that in bogs protected from pigs by fencing, there has been excellent recovery of native species. One area, Greensword Bog, was so severely damaged that there was little vegetation of any type remaining with over 90 percent of the area pig-plowed bare ground. Following fencing of Greensword Bog, the coverage of native plants increased from 6 percent to 90 percent in four years. By the late 1980s, three additional fenced exclosures were constructed protecting the park's four largest montane bogs.

ARGYROXIPHIUM VIRESCENS

Sunflower family (Asteraceae)

The pasture lands along the road that winds up to the park (Route 378) were once covered by thick, native forests of *koa*, *'ōhi'a*, and other trees 50-100 feet tall. By the 1900s, this forest had been basically destroyed through burning, logging, and cattle grazing. Today, the small groves of tall trees in these pasture lands are planted blue gum *(Eucalyptus globulus)* and swamp mahogany *(Eucalyptus robustus)*, both from Australia. It was in this area that another silversword relative, *Argyroxiphium virescens*, once flourished just above the dense rain forest at perhaps 6000-7000 feet elevation.

Argyroxiphium virescens was apparently not uncommon in the mid-1800s. When conifer and eucalyptus trees were planted in 1910 in an area later known as Hosmer Grove, several populations of this species still survived nearby. These were, however, threatened by the browsing of cattle and feral goats and after 1945, this species was thought to be extinct. However, a small population of the species was discovered and protected in the late 1980s. As of 1992, this silversword relative may again be on the verge of extinction, known only from a single plant in the wild. Propagation of this species by tissue culture is being attempted and though initial progress is hopeful, the final result is far from certain.

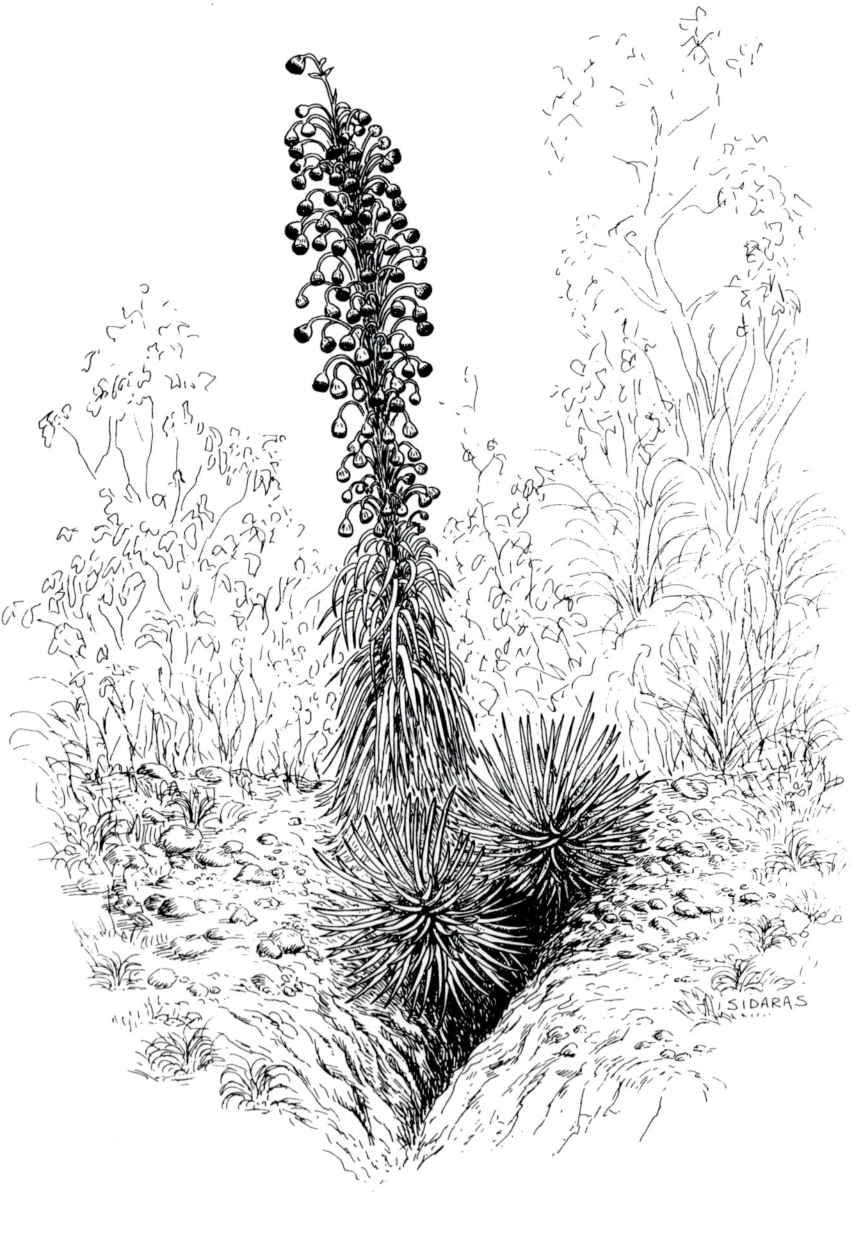

ARGYROXIPHIUM VIRESCENS
Sunflower family (Asteraceae)

HAWAIIAN LOBELIAS

'Ōhā, hāhā and other Hawaiian names
Clermontia tuberculata,
Clermontia samuelii, and
Lobelia gloria-montis illustrated
Lobelia family (Campanulaceae)

Of all Hawaiian plants, the lobelias represent the most prolific speciation from a single plant ancestor. At least four endemic Hawaiian genera, *Clermontia, Cyanea, Rollandia,* and *Delissia,* totalling just under one hundred species, all evolved from a common ancestor that arrived in Hawai'i millions of years ago.

Since the 1800s, most populations of Hawaiian lobelias have declined and dozens of species have been lost forever. Extinction and massive destruction of lobelia habitat by cattle were noted by botanical explorer J. F. Rock and by other early twentieth-century botanists. More recently, feral pigs and alien plant invasions have had obvious impacts, but other as yet poorly understood factors may also have played important roles. The dwindling of endemic honeycreeper and honeyeater birds (Meliphagidae), important in pollination of many endemic lobelias, may have further reduced the abundance of many species. Most of these pollinating birds have been extirpated or have highly depleted populations.

Haleakalā National Park has about twenty species of *Clermontia* and *Cyanea*, primarily in rain forests. Some species formerly present in the park have been eliminated by cattle and feral goats and pigs. The last individuals of *Cyanea pohaku*, a small soft-wooded tree of the Haleakalā shrubland, were destroyed by cattle by 1919, only nine years after J. F. Rock had discovered the last surviving population on Pu'u Nianiau, not far from the current park headquarters. This very distinctive species, known only from the collections of one scientist on one cinder cone, is now extinct.

In the high-elevation rain forests of northeastern Haleakalā at 5000-6500 feet elevation, *Clermontia tuberculata* is still relatively common. *Clermontia samuelii* is much rarer, known from few (fewer than 50) individuals. Feral pigs reached these upper rain forests only since the 1970s, and as yet their numbers are relatively low. As the lobelias are favored by pigs as food, protection from feral pigs is crucial for their long-term survival.

Cyanea copelandii subsp. *haleakalaensis* is found only on East Maui at two rain forest sites, one near Olinda and the other in the park's Kīpahulu Valley. This orange-yellow flowered species survives in moderate numbers on steep, rocky walls in the lower level of Kīpahulu Valley at an elevation of 3000-3500 feet. These areas are inaccessible to pigs and free of the dense growth of the alien Hilo grass *(Paspalum conjugatum)* which covers much of the nearby forest understory. The very rare *kuenui, Cyanea grimesiana*, had last been seen in the park in the 1920s until a population was recently rediscovered. The newly found population of this rare lobelia with pinnately-divided leaves, resembling tree fern leaves, contains far more individuals than are known from all other populations statewide, but is threatened by feral pigs and invasive weeds.

The genus *Lobelia* is represented in the park by three species found in rain forests and moist upper-elevation shrublands. *Lobelia grayana* and *Lobelia hillebrandii* both have beautiful purple-blue flowers. The first species is found at higher elevations above the forest, while the latter occurs only in rain forest. *Lobelia gloria-montis*, a plant of bog margins, is one of the most spectacular native plants in the Hawaiian archipelago. This plant, which resembles the giant lobelias of the mountains of East

Africa, has giant rosettes as wide as a human arm span that sit atop pole-like stems over 10 feet tall. The flowering plants reach 15 feet in height, topped by a spectacular candelabra of creamy white flowers. With this height, the majestic flowering plants may sometimes be visible for miles. Like the silversword, this giant lobelia flowers only once, then dies.

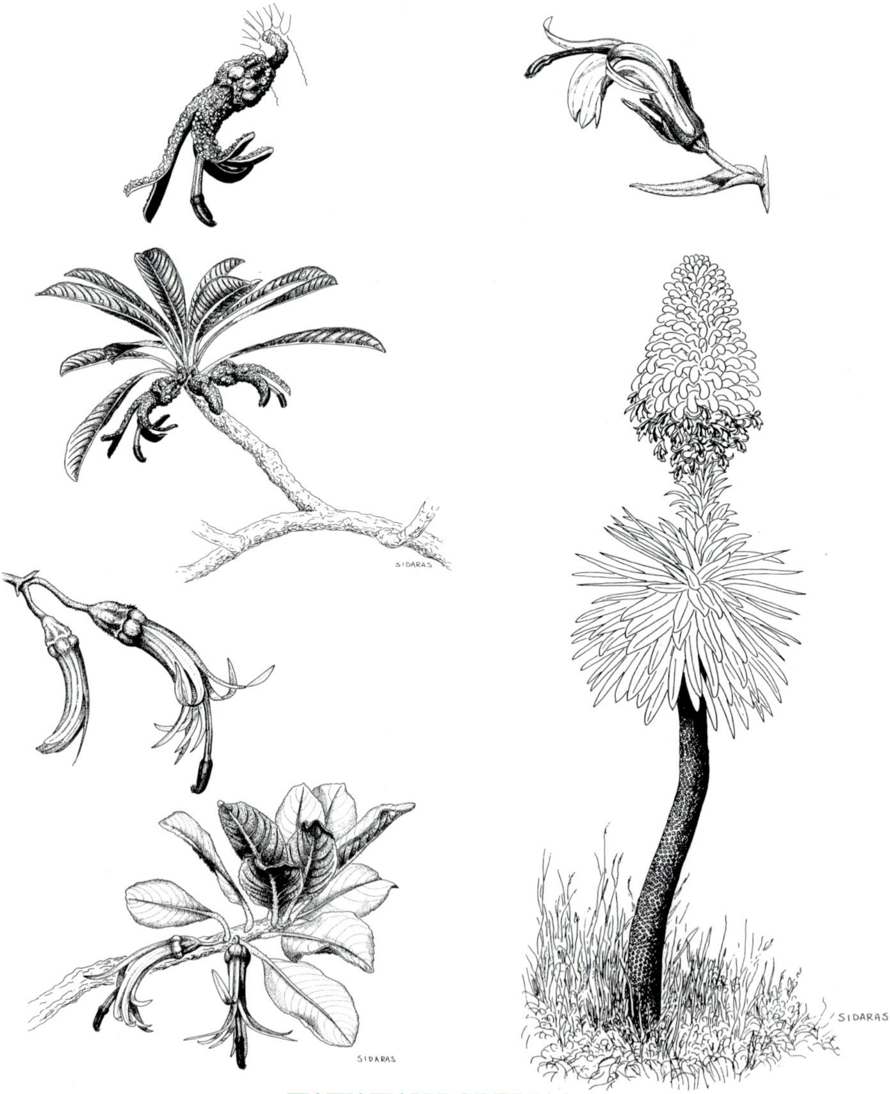

HAWAIIAN LOBELIAS

'Ōhā, hāhā, and other Hawaiian names

Clermontia tuberculata (top left), *Clermontia samuelii* (bottom left), and *Lobelia gloria-montis* (right)

Lobelia family (Campanulaceae)

NOHO-ANU
Red-flowered tree geranium
Geranium arboreum
Geranium family (Geraniaceae)

The red-flowered tree geranium is endemic to Haleakalā at 6000-7000 feet elevation in upper forests and shrublands. It is a weak-branched shrub up to 12 feet tall with whitish bark and bright-green leaves. These leaves are quite unlike those of most other native geraniums of Haleakalā, which are covered with silver hairs. This tall shrub typically occurs along stream gulches and banks, growing in thickets of dense vegetation such as *'ākala*, the native raspberry. The curved, irregular-shaped flower of the noho-anu is unique among the world's three hundred species of *Geranium*. These unique flowers apparently are adapted for pollination by native flower-feeding birds, and birds such as the *'apapane* can be observed visiting geraniums and probing the flowers for nectar with their long, curved bills.

Geranium arboreum was first collected in 1841. Once probably a common shrub, the total remaining population of this species has been reduced to several hundred individuals by the browsing of cattle and goats. Otto Degener, in his *Plants of Hawaii National Parks* published in 1930 stated that "this plant grows within the cattle ranges and therefore may be exterminated any day." *Geranium arboreum* has managed to survive into the 1990s, although establishment of new seedlings occurs only rarely.

Geranium arboreum was known to the Hawaiians as *noho-anu* (literally translated as "cold-dwelling") in reference to its cool, upper-elevation distribution. The *noho-anu* is found primarily in gulches on the western slopes of Haleakalā, with most populations outside the park. The largest number of surviving individuals occur within Kula Forest Reserve on the southwestern slope of the mountain. In 1983 the Maui Division of Forestry, with volunteer support, constructed a small fenced exclosure in this area to protect one of the larger populations of the shrub from destruction by cattle and pigs.

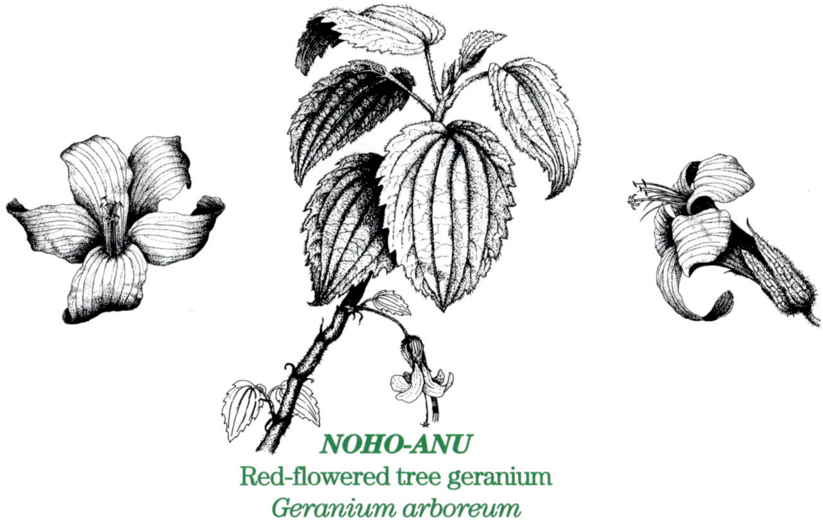

NOHO-ANU
Red-flowered tree geranium
Geranium arboreum
Geranium family (Geraniaceae)

ALE
Plantago princeps
Plantain family (Plantaginaceae)

ALE

Plantago princeps

Plantain family (Plantaginaceae)

Ale, or *Plantago princeps*, is one example of the many native species so severely threatened by feral goats that they now exist only in relatively inaccessible areas such as vertical cliff faces. At present, only three small populations of this species are known in the park. On the other Hawaiian islands, *Plantago princeps* now survives at precariously low population levels. In many areas where it was once common, such as Nu‘uanu Pali on O‘ahu, it no longer occurs at all.

Since feral goats are now under control within Haleakalā National Park, depleted populations of *ale* may eventually increase in numbers. However, the extremely reduced population levels may make reproduction difficult, especially if cross-pollination is necessary to produce fertile seed, a question as yet unanswered by biologists.

'ILIAHI

Haleakalā sandalwood
Santalum haleakalae
Sandalwood family (Santalaceae)

Only six hundred to eight hundred individuals of the rare Haleakalā sandalwood occur, mostly between 6000 and 8700 feet elevation. Sandalwoods are partially parasitic plants whose roots extract nutrients from the roots of neighboring plants. The Haleakalā sandalwood is a shrub or small tree that grows up to about 15 feet in height and bears clusters of bright-red flowers.

Sandalwood fruits are heavily preyed upon as food by the black rat *(Rattus rattus)* and the house mouse *(Mus domesticus)*, both introduced to Hawai'i in the 1800s. The fleshy, red-purple fruits appear adapted for bird dispersal, but are not eaten by any of the existing native birds. In former times, a now extinct, large Hawaiian bird may have eaten, scarified, and dispersed sandalwood seeds. Introduced game birds such as ring-necked pheasants and chukar partridges may occasionally aid the reproduction of this native species by passing the hard-to-germinate seeds through their digestive tracts. Out of the population of hundreds of Haleakalā sandalwoods, few trees have seedlings beneath them. Most of the trees with seedlings are located in Haleakalā crater in areas of very sparse vegetation. Trapping for rodents under these trees suggests extremely low rodent populations.

The Haleakalā sandalwood is capable of reproducing vegetatively, sprouting new growth from roots near the surface. This trait accounts for the "new" plants that can sprout 50 feet or more from a "parent" tree. Where protected from goats, these shoots become large bushes and, eventually, perhaps, "new" trees. New techniques, modeled after those used in germinating other species of sandalwood, have recently been successful with native Hawaiian species, including the Haleakalā sandalwood. Perhaps these techniques coupled with a carefully planned planting program will assure the long-term survival of this rare tree.

'ILIAHI
Haleakalā sandalwood
Santalum haleakalae
Sandalwood family (Santalaceae)

MAKOU
Hawaiian buttercup
Ranunculus mauiensis
Buttercup family (Ranunculaceae)

MAKOU
Hawaiian buttercups
Ranunculus mauiensis illustrated and *Ranunculus hawaiensis*
Buttercup family (Ranunculaceae)

The Hawaiian buttercups are fleshy herbs with broad, divided, often fuzzy leaves, and 1/2- to 1-inch wide bright-yellow flowers. Two species of endemic Hawaiian buttercups were fairly common on Haleakalā at 5000 to 9500 feet elevation until the beginning of this century. After the 1930s, buttercups became very rare, disappearing entirely from many areas. In 1927, botanist Otto Degener noted that *Ranunculus hawaiensis* was "relatively common" north of Hōlua, but in 1959, he could find only two plants. As of 1991, *Ranunculus hawaiensis* is apparently extirpated on Maui, and *R. mauiensis* is extremely rare, found in only a few remnant populations outside the park.

The destruction of the *koa* forests on the western slopes of Haleakalā by cattle, goats and pigs has undoubtedly caused the loss of many native plants, including Hawaiian buttercups. However, another event seems to have played a major role in the dramatic decline of this genus in Hawai'i in the past 60 years: the inadvertent introduction of certain mollusks, such as common garden slugs. A seedling of *Ranunculus mauiensis*, recently cultivated near park headquarters, was defoliated by non-native slugs *(Milax gigates)*. After the slugs were controlled, the young Maui buttercup recovered, matured, and flowered. The damage caused by slugs may have been a significant factor in the decline of the Hawaiian buttercups as well as some other rare Hawaiian plant species.

HAWAIIAN MINTS

Stenogyne haliakalae and
Stenogyne microphylla illustrated
Mint family (Lamiaceae)

There are four genera of Hawaiian mints, three of which are endemic. The two groups of forest mints, *Stenogyne* and *Phyllostegia*, have many species, some of which are found within Haleakalā National Park. Without aromatic foliage to deter browsing, Hawaiian mints have proved highly vulnerable to extinction after the introduction of cattle, pigs, and goats.

Of eight species of the endemic Hawaiian genus *Stenogyne* from East Maui, five are believed to be extinct. One Haleakalā *Stenogyne* is *S. haliakalae* (the species name a misspelling of the mountain's name). It was often collected on Haleakalā by naturalists prior to the 1920s, but has not been seen in recent years; it is likely extinct. The plant was reported as a stout vine with one inch long rose-pink blossoms that climbed high into the canopies of *'ōhi'a* and *'a'ali'i* trees. This rare Haleakalā mint was collected nine times

HAWAIIAN MINT
Stenogyne haliakalae
Mint family (Lamiaceae)

HAWAIIAN MINT
Stenogyne microphylla
Mint family (Lamiaceae)

between 1868 and 1920 and twice since then. The last collection of *Stenogyne haliakalae* in the park was made in 1937 at 5600 feet elevation in dryland forest in eastern Kaupō Gap, near the present trail. A collection made in 1978 in Kula Forest Reserve above ʻUlupalakua may have been from the last individual of the species.

Stenogyne microphylla is a climbing vine which once was probably much more common in the shrublands of upper Haleakalā from 5000 to 9000 feet elevation. This rare mint now survives only in the canopies of small trees, such as *māmane* and Haleakalā sandalwood, which afford some protection from feral goats. In the *māmane* forests within Haleakalā crater where goats have been fenced out, *Stenogyne microphylla* has increased slightly, but competition from weeds prevents full recovery.

Stenogyne kamehamehae is found in rain forests of Molokaʻi and Maui. This species has striking clusters of 3-inch white or red flowers. *Stenogyne rotundifolia*, a Maui endemic, has 1-inch pure white flowers and, though quite rare, still survives in upper elevation Haleakalā rain forests not degraded by feral pigs.

HAWAIIAN ORCHIDS
Platanthera holochila illustrated
Orchid family (Orchidaceae)

Hawai'i is famous for cultivated orchids. Over one thousand species have been introduced from tropical areas throughout the world, but the archipelago itself has only three native species. The lack of the right insect pollinator and the absence of proper symbiotic soil fungi in Hawai'i have been suggested as possible reasons for the limited numbers of Hawaiian orchid species.

Two of the three native orchids, *Anoectochilus sandvicensis* and *Liparis hawaiiensis*, are uncommon to rare in the rain forests of Haleakalā National Park. Both are small, fairly inconspicuous plants with spikes of pale, greenish flowers. The third species of native Hawaiian orchid, *Platanthera holochila*, was once fairly common in parts of Hawai'i, but has recently undergone a rapid decline. The only record of this orchid in the park was made by botanical explorer C.N. Forbes, who collected nine specimens in the rain forest near Wai'ānapanapa lake in 1920. Currently, only five or six small populations are known in the Hawaiian Islands. One of these is on East Maui in shrubland at 6600 feet elevation, just outside the boundaries of the park on land managed by The Nature Conservancy. Feral pig rootings near the orchids led to the construction of a small fence in 1984 to protect these rare plants. Some possible reasons for the abrupt decline of this species include damage by feral pigs, herbivory by introduced mollusks, or the introduction into Hawai'i in the late 1800s to early 1900s of an unknown orchid pathogen. Whatever the cause, the species may be close to disappearing forever.

HAWAIIAN ORCHID
Platanthera holochila
Orchid family (Orchidaceae)

VII. CREDITS

The late Wayne Gagné was a crucial source of much knowledge and inspiration for this project and our work in general. We dedicate this project to him.

We would like to acknowledge the exceptional work of artist Nanci Sidaras. The subjects of her line drawings include many plants and animals rarely seen and never before illustrated. We also thank artist Arthur Kodani who drew the flightless lacewing *Pseudopsectra lobipennis*.

This booklet was greatly improved by the input of Dorothy Barrère, Robert Butterfield, Charles Chimera, Sheila Conant, Clif Davis, Kathy English, Pam Frierson, Betsy Gagné, Rosie Gillespie, William Gilmartin, Vikki Greive, Robert Hobdy, Cathleen Hodges, Jim Jacobi, Denise Koon, William Mull, Eric Nishibayashi, Thane Pratt, Pamela Rasfeld, Kim Sikoryak, Charles Stone, and Philip Thomas.

We would like to thank the following photographers for the use of their excellent images: Betsy Gagné (p. 27 silversword "family"; silversword flower head; p. 28, lobelia spikes; p. 30, dragonfly), Jack Jeffrey (p. 29 and back cover, crested honeycreeper), James Larson (p. 27, petrel chick), David Muench (front and back cover, western Ko'olau Gap), and William Mull (p. 29 and back cover, happy-face spider). The remainder of the photographs were taken by Art Medeiros.

The book was designed by Paul Martin and printed by Tongg Publishing Company.

VIII. GLOSSARY

Scientific terms used in the text are defined below in the context of their use in this booklet. The words shown in all capital letters are defined elsewhere in the glossary.

ADAPTIVE RADIATION. Evolutionary divergence of SPECIES originating from a single POPULATION of ancestors into a variety of different physical forms, usually prompted by diversification in the use of resources or HABITATS. Classic examples in Hawai'i include the silverswords, LOBELIAS, HONEYCREEPERS, and vinegar flies.

ALIEN. Not NATIVE; introduced from another geographic area by humans, not necessarily purposefully.

ARCHIPELAGO. A group of islands.

ARTHROPOD. Belonging to the phylum Arthropoda, a group of invertebrate animals with jointed body and limbs that includes insects, spiders, scorpions, mites, centipedes, etc.

BIOMASS. The total amount (weight) of living matter. The term may be used in reference to an individual, a SPECIES, or an ECOSYSTEM.

BIOSPHERE RESERVE. A NATURAL AREA with internationally significant biological resources, recognized by the United Nations Educational, Scientific and Cultural Organization (UNESCO) as part of an international network of such reserves devoted to biological preservation, research, and education.

CANOPY. The uppermost, spreading portion of a forest which receives the maximum amount of radiation from the sun.

COEVOLUTION. EVOLUTION of two or more organisms, e.g. a plant and a

POLLINATOR, specifically and reciprocally in response to each other.

COMMUNITY. An interacting group of POPULATIONS of plant and animal SPECIES occupying a given area.

ECOSYSTEM. The complex of a biological COMMUNITY and its environment functioning as an ecological unit.

ENDANGERED SPECIES. An animal or plant SPECIES formally designated (in the *Federal Register*) as "endangered" by the U. S. Fish and Wildlife Service, within the framework of the Endangered Species Act of 1973.

ENDEMIC. Found naturally in only a specified region or locality. For example, *Santalum haleakalae*, the Haleakalā sandalwood, grows on only the upper slopes of Haleakalā and is therefore endemic to that area. In this book, when endemic is used without specifying the area, it is assumed that the SPECIES is endemic to the Hawaiian Islands.

EVOLUTION. The development over time of a biological group.

EXTINCT. No longer existing. Usually applied to an entire SPECIES or SUBSPECIES.

EXTIRPATED. Destroyed completely and no longer existing. Usually applied to a local POPULATION of a SPECIES.

FAMILY. A group of related plants or animals forming a category ranking above a GENUS and below an order and usually comprising several to many GENERA.

FERAL. Not domesticated or cultivated, but escaped from domestication and living in the wild.

GALÁPAGOS FINCHES. A group of birds ENDEMIC to the Galápagos Islands that comprise the most widely known example of evolutionary ADAPTIVE RADIATION, because of their association with Charles Darwin and his theory of EVOLUTION.

GENUS (pl. GENERA). A category of biological classification, ranking between FAMILY and SPECIES, comprising related species or a single species with unusual differentiation; designated by a capitalized Latin name. Using the silversword group as an example, *Argyroxiphium* is a genus in the aster family (Asteraceae) ENDEMIC to the Hawaiian Islands and comprised of five closely related species, three of which *(A. sandwicense, A. grayanum, and A. virescens)* are illustrated and discussed in this booklet.

HABITAT. The place or type of site where a plant or animal normally lives and grows.

HERBIVORE. A plant-eating animal.

HERBIVORY. The act or result of an animal feeding on a plant.

HONEYCREEPER. Any member of an ENDEMIC subfamily (Drepanidinae) of cardueline finches (Fringillidae) derived from a single colonization of the Hawaiian Islands by a finch-like ancestor from the Americas.

HONEYEATER. Any member of a large FAMILY (Meliphagidae) of predominantly nectar-feeding birds found throughout Australia and many Pacific islands, acting in many cases as important POLLINATORS of plants and having mournful, bell-like calls.

INDIGENOUS. Growing and living naturally in a particular locality. In the Hawaiian Islands the term is often used to designate SPECIES which are NATIVE but not ENDEMIC. For example, the Hawaiian short-eared owl or *pueo (Asio flammeus)* is INDIGENOUS at the species level, since the same species occurs on the North American mainland.

INTRODUCED. Not NATIVE. Brought from another geographic area by humans.

KOA. One of the most common NATIVE forest trees, *Acacia koa* supports a tremendous variety of associated plant, insect, and bird life. Large tracts of *koa* forests which formerly occupied the flanks of Haleakalā between Makawao and 'Ulupalakua have disappeared with barely a trace, due to the effects of fire and grazing by cattle.

LARVA (pl. LARVAE). The wingless and often grub-like immature form of many insects; the larva hatches from the egg, grows while passing through several molts (shedding of an outer layer), and is transformed into a PUPA from which the adult later emerges.

LEEWARD. The direction opposite that from which the predominant wind blows. In the Hawaiian Islands, refers to the southwestern side of islands, where precipitation, driven by northeast trade winds, is lowest. The dry side of a Hawaiian island. See also WINDWARD.

LOBELIA. Generally refers to members of the plant FAMILY Lobeliaceae which in Hawai'i has evolved into many SPECIES contained in six GENERA (*Brighamia, Clermontia, Cyanea, Delissia, Lobelia,* and *Rollandia*). When capitalized and underlined or italicized (i.e., *Lobelia*), it refers to a genus contained within family Campanulaceae.

MALARIA (AVIAN). A debilitating and deadly disease of birds, especially NATIVE Hawaiian birds, caused by sporozoan parasites (GENUS *Plasmodium*) in the red blood cells and transmitted by the bite of ALIEN mosquitoes. Closely related to a similar disease of humans that is characterized by periodic attacks of chills and fever and is also transmitted by mosquitoes but absent from the Hawaiian Islands.

MOLLUSKS (or MOLLUSCS). Any of a large phylum (Mollusca) of invertebrate animals (e.g. slugs, snails, and clams) with a soft, unsegmented body usually enclosed in a calcareous shell.

NATIONAL PARK. An area of Federal land designated by an act of Congress as having special scientific, historical, and/or scenic significance, and administered and protected by the U. S. Department of the Interior, National Park Service.

NATIVE. Growing and living naturally in a particular locality. In the Hawaiian Islands, the term is generally used to designate SPECIES that are either ENDEMIC to the Hawaiian Islands, or INDIGENOUS but not endemic.

NATURAL AREA. An area designated for preservation of original native COMMUNITIES.

NATURAL AREA RESERVE SYSTEM (NARS). A system of legally protected NATURAL AREAS owned by the State of Hawai'i and administered by the Department of Land and Natural Resources.

OCEANIC ISLAND. An island, generally of volcanic origin, located in the deep ocean far beyond the continental shelf. Contrast with continental island.

'ŌHI'A. The Hawaiian name for the most common of NATIVE forest trees, *Metrosideros polymorpha,* which has beautiful red (or more rarely yellow) pincushion flowers (called *lehua*). The nectar of these flowers provides native forest birds with much of their food supply.

PERKINS, R.C.L. (1866-1955). An English naturalist. Perkins' early work with Hawaiian birds and insects constitutes, in many cases, the bulk of current knowledge of many NATIVE SPECIES. As a result of his energetic Hawaiian field work (1892-1901), Perkins is the only Westerner to have seen and recorded the habits and HABITATS of many native birds and insects that have disappeared from their former haunts.

POLLINATOR. An agent (usually an insect, but in Hawai'i sometimes a bird) that transfers pollen to the stigma of a flower, accomplishing cross-fertilization.

POPULATION. A group of potentially interbreeding organisms of a SPECIES occurring within a defined geographical area.

POX (AVIAN). The common name for this slow-developing but potentially fatal viral disease that affects a great variety of bird SPECIES; mosquito bites are the most common form of infection of birds by POX. Native Hawaiian birds seem especially vulnerable to avian pox.

PREDATOR. An animal that obtains food by killing and consuming other animals.

PRISTINE. A term often used in conservation terminology to describe a HABITAT largely unmodified or undisturbed from its original condition.

PUPA (pl. PUPAE). An intermediate stage in the development of an immature insect from LARVA to adult.

RAIN FOREST. Forest, usually in a tropical area, that receives heavy rainfall throughout the year, sometimes defined as averaging greater than 100 inches per year, with no less than 4 inches in any one month.

ROCK, J. F. (1884-1962). A world-renowned botanist, plant collector, and explorer who left his native Austria to seek a warm, tropical climate in which to recover from tuberculosis. Rock arrived in Hawai'i in 1907 and spent about twelve years collecting and publishing descriptions of the NATIVE Hawaiian plants before leaving to explore China. He is generally regarded as one of the finest of Hawaiian botanists and his well-written books with their excellent black-and-white photographs are considered classics.

SPECIES. A category of biological classification ranking below GENUS; designated by a lower case Latin name. The term describes a group of individuals with common attributes that are capable also of interbreeding. See also GENUS.

SUBSPECIES. A category of biological classification ranking below SPECIES. One or more subspecies are sometimes designated within a species, when geographically separated POPULATIONS have some locally distinct feature. For example, the Haleakalā silversword (*Argyroxiphium sandwicense* subsp. *macrocephalum*) and Mauna Kea silversword (*Argyroxiphium sandwicense* subsp. *sandwicense*) are considered distinct at the subspecies level; they belong to the same species, *Argyroxiphium sandwicense*.

SYMBIOTIC. Participation of two dissimilar organisms in a close association that results in mutual benefits.

UNDERSTORY. The layer of low-growing vegetation established at and just above ground level in a forest.

WINDWARD. The direction from which the predominant wind blows. In the Hawaiian Islands, refers to the northeastern side of islands, where precipitation, driven by northeast trade winds, is heaviest. The wet side of a Hawaiian island. See also LEEWARD.

IX. ANNOTATED RECOMMENDATIONS FOR FURTHER READING

HAWAIIAN VOLCANISM

Decker, R., and B. Decker. 1991. *Volcano Watching*. Hawai'i Natural History Association. Hawai'i Volcanoes National Park, Hawai'i. An affordable, readable yet accurate guide to Hawaiian volcanoes as well as selected aspects of volcanism in general.

Macdonald, G.A., A.T. Abbott, and F.L. Peterson. 1983. *Volcanoes in the Sea: the Geology of Hawai'i*. University of Hawai'i Press, Honolulu. An authoritative and readable textbook on the geology of the state. Readily available, this is probably the best single reference on the subject.

GENERAL HAWAIIAN NATURAL HISTORY

Bioscience. April 1988. Special issue on Hawaiian biology. A fine, fairly recent overview.

Carlquist, S. 1980. *Hawai'i: a Natural History*. 2nd ed. Pacific Tropical Botanical Garden, Lāwa'i, Kaua'i. An excellent introduction to Hawai'i, with many illustrations.

Daws, G. 1988. *Hawai'i: The Islands of Life*. Signature Publishing, Honolulu. Produced by The Nature Conservancy of Hawai'i with vivid text and breathtaking photographs, this is the ultimate coffee table book on the natural history of Hawai'i.

Gagné, W.C. 1975. Hawai'i's tragic dismemberment. *Defenders* 50:461-469. A classic assessment of the gravity of the ongoing loss of the natural heritage of Hawai'i. Dr. Wayne Gagné (1942-1988) was an entomologist, educator, and conservation activist of boundless energy and ability.

Trends in Ecology and Evolution. 1987. Special Issue: Hawaiian Evolutionary Biology. 2(7): 175-229. Provides concise reviews of fairly recent and ongoing research, highly recommended for overviews of the various taxonomic groups.

Stone, C.P., and D.B. Stone, eds. 1989. *Conservation Biology in Hawai'i*. University of Hawai'i Cooperative National Park Resources Studies Unit. Distributed by University of Hawai'i Press, Honolulu. Inexpensive, up-to-date, readable, and authoritative: a must for anyone interested in the topic.

Zimmerman, E.C. 1948. *Insects of Hawai'i, Vol. 1. Introduction*. University of Hawai'i Press, Honolulu. An older work, out of print, hard to get, yet still a superb introduction to evolution in the Hawaiian Islands.

PLANTS AND PLANT ECOLOGY

Carr, G.D. 1985. Monograph of the Hawaiian Madiinae (Asteraceae): *Argyroxiphium*, *Dubautia*, and *Wilkesia*. *Allertonia* 4: 1-123. A fine modern taxonomic treatment of the twenty-eight species of the silversword alliance.

Degener, O., and I. Degener. 1932-1980. *Flora Hawaiiensis or the New Illustrated Flora of the Hawaiian Islands*.

Privately published. Though too technical for most readers, this work provides an indispensable source of botanical information.

Degener, O. 1973. *Plants of Hawai'i National Parks Illustrative of Plants and Customs of the South Seas.* Braun-Brumfield, Inc., Ann Arbor. Somewhat out-of-date, but inexpensive and with much good first-hand information.

Kepler, A.K. 1983. *Hawaiian Heritage Plants.* The Oriental Publishing Co., Honolulu. Good color photographs of Hawaiian flora, along with information on history, ethnobotany, and natural history.

Kepler, C.B., and A.K. Kepler. 1988. *Haleakalā: A Guide to the Mountain.* Mutual Publishing Co., Honolulu. A well-illustrated, informative guide to the landscapes of East Maui, including Haleakalā National Park.

Lamb, S.H. 1981. *Native Trees and Shrubs of the Hawaiian Islands.* Sunstone Press, New Mexico. Recent work in a subject where few other references are readily available.

Lamoureux, C.H. 1976. *Trailside Plants of Hawai'i's National Parks.* Hawai'i Natural History Association, Hawai'i Volcanoes National Park, Hawai'i. An excellent, inexpensive, well-illustrated guide to common plant species in the parks, especially useful for Hawai'i Volcanoes National Park. Highly recommended.

Mueller-Dombois, D. 1987. Natural dieback in forests. *Bioscience* 37: 575-583. A good introduction to the author's voluminous work on the phenomenon of *'ōhi'a* dieback in Hawai'i and its relation to forest dynamics elsewhere.

Rock, J.F. [1913] 1974. *The Indigenous Trees of the Hawaiian Islands.* Pacific Tropical Botanical Garden, Lāwa'i, Kaua'i. A well-written work with beautiful and historic black and white photographs still much in use today; a classic.

St. John, H. 1973. *List and Summary of the Flowering Plants in the Hawaiian Islands.* Memoir no. 1. Pacific Tropical Botanical Garden, Lāwa'i, Kaua'i. A landmark work at its publication, this checklist continues to be a valuable reference.

Sohmer, S.H., and R. Gustafson. 1987. *Plants and Flowers of Hawai'i.* University of Hawai'i Press, Honolulu. Good information and beautiful photographs are well-presented.

Vitousek, P.M. and others. 1987. Biological invasion by *Myrica faya* alters ecosystem development in Hawai'i. *Science* 238: 802-804. A excellent but somewhat technical example of fairly recent and ongoing ecological work in Hawai'i.

Wagner, W., D.L. Herbst, and S.H. Sohmer. 1990. *Manual of Flowering Plants of Hawai'i.* University of Hawai'i Press, Honolulu. This important work satisfies a long-standing need for an up-to-date, relatively comprehensive manual.

Whiteaker, L.D. 1983. The vegetation and environment in the Crater District of Haleakala National Park. *Pacific Science* 37: 1-24. An good but technical treatment of the topic.

VERTEBRATE FAUNA (BIRDS AND MAMMALS)

Banko, P.C., and D.A. Manuwal. 1982. *Life History, Ecology, and Management of Nēnē (Branta sandvicensis) in Hawai'i Volcanoes and Haleakala National Parks.* University of Washington Cooperative National Park Resources Studies Unit, Seattle. Gives most of the available information on the Hawaiian goose in Haleakalā National Park.

Berger, A.J. 1981. *Hawaiian Birdlife.* 2nd ed. University of Hawai'i Press, Honolulu. An important general reference on the birds of Hawai'i.

Carothers, J.H., S.R. Sabo, and R.B. Hansen. 1983. Ecological observations on an endangered species: the Maui Parrotbill, *Pseudonestor xanthophrys*. *American Birds* 37(5): 820-821. A technical article that contains the most thorough documentation available on this rare and unique Maui honeycreeper.

Conant, S., and M.S. Kjargaard. 1985. Annotated checklist of birds of Haleakalā National Park, Maui, Hawai'i. *Western Birds* 15: 97-110. A list of native and introduced birds of the park, with notes on bird distribution and ecology.

'Elepaio, Vols. 1-53. This journal of the Hawai'i Audubon Society is inexpensive and has timely and concise articles dealing with Hawaiian birds and conservation issues affecting them. Back issues are available.

Hawai'i Audubon Society. 1989. *Hawai'i's Birds*. Hawai'i Audubon Society, Honolulu. An excellent inexpensive guide. Highly recommended.

James, H.F., and S.L. Olson. 1991. *Descriptions of Thirty-two New Species of Birds from the Hawaiian Islands: Part II. Passeriformes.* Ornithological Monographs No. 46, The American Ornithologists' Union, Washington D.C. The technical descriptions of 14 new species of honeycreepers and two new species of crows based on fossil and semi-fossilized bird bones found in lava tubes, sinkholes, and sand dunes.

Kear, J., and A.J. Berger. 1980. *The Hawaiian Goose: An Experiment in Conservation.* Buteo Books, Vermillion, S.D. Gives the biology of the *nēnē* and the history of conservation efforts.

Munro, G.C. 1944. *Birds of Hawai'i.* Tongg Publ. Co., Honolulu. Though long out of print, difficult to obtain, and somewhat out-of-date, an important work with much first-hand information about native birds.

Olson, S.L., and H.F. James. 1982. *Prodromus of the Fossil Avifauna of the Hawaiian Islands.* Smithsonian Contributions to Zoology, no. 369. Smithsonian Institution, Washington D.C. The first major report on the ongoing work that is revolutionizing knowledge of the Hawaiian bird fauna.

Olson, S.L., and H.L. James. 1991. *Descriptions of Thirty-two New Species of Birds from the Hawaiian Islands: Part I. Non-Passeriformes.* Ornithological Monographs No. 45, The American Ornithologists' Union, Washington D.C. The technical descriptions of new petrel, ibis, rails, owls, hawk, eagle, geese, and goose-like birds based on fossil and semi-fossilized bird bones found in lava tubes, sinkholes, and sand dunes.

Pratt, H.D., P.L. Bruner, and D.G. Berrett. 1987. *A Field Guide to the Birds of Hawai'i and the Tropical Pacific.* Princeton Univ. Press, Princeton, N.J. Provides a good perspective on the uniqueness of the Hawaiian bird fauna through contrast with the fauna of other Pacific archipelagoes.

Scott, J.M. and others. 1986. *Forest Bird Communities of the Hawaiian Islands: Their Dynamics, Ecology and Conservation.* Studies in Avian Biology No. 9. Cooper Ornithological Society. Allen Press, Lawrence, Kansas. A comprehensive very useful work for those interested in Hawaiian birds, this volume summarizes the results of the 1980 Maui forest-bird survey.

Simons, T.R. 1985. Biology and behavior of the endangered Hawaiian Dark-Rumped Petrel. *The Condor* 87: 229-245. The definitive work to date on this rare Hawaiian seabird that nests in cliffs of upper Haleakalā.

Tomich, P.Q. 1986. *Mammals in Hawai'i: a Synopsis and Notational Bibliography.* 2nd ed. B.P. Bishop Museum Special Publication no. 76. B.P.

Bishop Museum Press, Honolulu. A good review of information regarding the few native and many alien mammals.

van Riper, C. III and others. 1986. The epizootiology and ecological significance of malaria in Hawaiian land birds. *Ecological Monographs* 56: 327-344. The definitive work to date on effects of disease on birds in Hawai'i. This important aspect of Hawaiian conservation biology has received little attention.

INVERTEBRATE FAUNA

Howarth, F.G. 1990. Hawaiian terrestrial arthropods: an overview. *Bishop Museum Occasional Papers* 30: 4-26. B.P. Bishop Museum Press, Honolulu. An authoritative, up-to-date synthesis embracing ecology, evolution, and conservation biology.

Howarth, F.G., and W.P. Mull. 1992. *Hawaiian Insects and Their Kin.* University of Hawai'i Press, Honolulu. Excellent pictures and text make this the finest popular work on the Hawaiian insects and other arthropods.

Perkins, R.C.L. 1913. Introduction. In *Fauna Hawaiiensis*, ed. D. Sharp, Vol. 1, xv-ccxxviii, plates 1-16. The University Press, Cambridge, U.K. One of the finest existing works on Hawaiian invertebrates. Unfortunately, the quality of the paper used was apparently poor as the few surviving copies of this classic are extremely brittle and fragile.

Zimmerman, E.C. and others. 1948-1981. *Insects of Hawai'i, Volumes 1-14.* University of Hawai'i Press, Honolulu. This important series provides valuable information and the technical taxonomy for all Hawaiian insects except the Coleoptera (beetles) and the Hymenoptera (ants, wasps, and bees).

EFFECTS OF HUMANS ON ENDEMIC HAWAIIAN BIOTA AND PRESCRIPTIONS FOR MANAGEMENT

Cuddihy, L.W., and C.P. Stone. 1990. *Alteration of Native Hawaiian Vegetation.* University of Hawai'i Cooperative National Park Resources Studies Unit. Distributed by University of Hawai'i Press, Honolulu. An extremely valuable synthesis of information not otherwise readily available.

Culliney, J.L. 1988. *Islands in a Far Sea: Nature and Man in Hawai'i.* Sierra Club Books, San Francisco. A good overview dealing with both terrestrial and marine issues.

Kirch, P.V. 1985. *Feathered Gods and Fishhooks: An Introduction to Hawaiian Archaeology and Prehistory.* University of Hawai'i Press, Honolulu. A highly readable synthesis of the archaeology of the Hawaiian Islands, the culture of the Hawaiians, and the consequences of their colonization.

Stone, C.P., and L.L. Loope. 1987. Effects of introduced animals on native species in Hawai'i: what has been done, what needs doing, and the role of national parks. *Environmental Conservation* 14: 245-258. Details the status of conservation efforts in Haleakalā and Hawai'i Volcanoes National Parks.

Stone, C.P., and J.M. Scott, eds. 1985. *Hawai'i's Terrestrial Ecosystems: Preservation and Management.* University of Hawai'i Cooperative National Park Resources Studies Unit. Distributed by University of Hawai'i Press, Honolulu. A valuable and authoritative compendium of papers covering a wide spectrum of issues in conservation of Hawaiian ecosystems.

Stone, C.P., C.W. Smith, and J.T. Tunison. 1992. *Alien Plant Invasions in Native Ecosystems of Hawai'i: Management and Research.* University of

Hawai'i Cooperative National Park Resources Studies Unit. University of Hawai'i Press, Honolulu. A comprehensive review of the invasion and management of alien plants, a crucial issue in the long-term conservation of native Hawaiian ecosystems.

EVOLUTION ON ISLANDS

Carlquist, S. 1974. *Island Biology.* Columbia University Press, New York. A review of the ideas of the author and others on the topic.

Grant, P.R. 1986. *Ecology and Evolution of Darwin's Finches.* Princeton University Press, Princeton, N.J. Provides abundant detail on the biology of one of the best-known island bird groups, as well as a fine introduction to the biology of the Galápagos archipelago.

Williamson, M. 1981. *Island Populations.* Oxford University Press, Oxford, U.K. A well-organized, readable account of the subject.

CONSERVATION BIOLOGY

Soulé, M.E., ed. 1986. *Conservation Biology: the Science of Scarcity and Diversity.* Sinauer Associates, Sunderland, Mass. This book provides a wealth of detail on the subject. Though the problems are daunting, the research under way shows promise of providing details necessary for an understanding of local needs and global concerns.

Wilson, E.O., ed. 1988. *Biodiversity.* National Academy Press, Washington D.C.. A fairly recent, authoritative, and highly readable treatment of extinction-related issues.

NOTES